TOXICANTS, HEALTH AND REGULATION SINCE 1945

Studies for the Society for the Social History of Medicine

Series Editors: David Cantor
Keir Waddington

Titles in this Series

1 Meat, Medicine and Human Health in the Twentieth Century
David Cantor, Christian Bonah and Matthias Dörries (eds)

2 Locating Health: Historical and Anthropological Investigations of Place and Health
Erika Dyck and Christopher Fletcher (eds)

3 Medicine in the Remote and Rural North, 1800–2000
J. T. H. Connor and Stephan Curtis (eds)

4 A Modern History of the Stomach: Gastric Illness, Medicine and British Society, 1800–1950
Ian Miller

5 War and the Militarization of British Army Medicine, 1793–1830
Catherine Kelly

6 Nervous Disease in Late Eighteenth-Century Britain: The Reality of a Fashionable Disorder
Heather R. Beatty

7 Desperate Housewives, Neuroses and the Domestic Environment, 1945–1970
Ali Haggett

8 Disabled Children: Contested Caring, 1850–1979
Anne Borsay and Pamela Dale (eds)

Forthcoming Titles

A Medical History of Skin: Scratching the Surface
Jonathan Reinarz and Kevin Siena

The Care of Older People: England and Japan, A Comparative Study
Mayumi Hayashi

TOXICANTS, HEALTH AND REGULATION SINCE 1945

EDITED BY

Soraya Boudia and Nathalie Jas

Routledge
Taylor & Francis Group
LONDON AND NEW YORK

First published 2013 by Pickering & Chatto (Publishers) Limited

Published 2016 by Routledge
2 Park Square, Milton Park, Abingdon, Oxfordshire OX14 4RN
711 Third Avenue, New York, NY 10017, USA

First issued in paperback 2015

Routledge is an imprint of the Taylor & Francis Group, an informa business

British Library Cataloguing in Publication Data

Toxicants, health and regulation since 1945. – (Studies for the Society for the Social History of Medicine)
1. Hazardous substances – Health aspects. 2. Hazardous substances – Law and legislation. 3. Hazardous substances – History – 20th century.
I. Series II. Boudia, Soraya. III. Jas, Nathalie.
363.1'7-dc23

ISBN-13: 978-1-138-66476-0 (pbk)
ISBN-13: 978-1-8489-3403-0 (hbk)

Typeset by Pickering & Chatto (Publishers) Limited

CONTENTS

ACKNOWLEDGEMENTS

As editors of the book, we are indebted to many people without whom this project would not have been possible. We are thankful for the many stimulating and friendly discussions we had with Francis Chateauraynaud, Josquin Debaz and the other participants of the Research Project 'FADO' – *Low Doses and Expertise: Historical and Sociological Approaches*. We also would like to thank Amy Dahan, Jean-Paul Gaudillière, Dominique Pestre and Sezin Topçu for the sometimes lively, but always exceedingly inspiring discussions we had had during the three years that the seminar on the *Global Government of Technosciences* (GATSEG) lasted. The debates we had within the FADO and GATSEG projects greatly improved the conception of the book and helped us strengthen our main thesis. The generous support of the Agence Nationale de Recherche (ANR) to these two projects facilitated our own research for this book.

The ideas presented in this book were worked out in significant respects in several workshops and seminars. We would like to thank the commentators of the earlier versions of the first chapter of the book. Beate Bächi, Christophe Bonneuil, Olivier Borraz, Yves Cohen, David Demortain, Jean-Baptiste Fressoz, Claude Gilbert, Emmanuel Henry, Jean-Noël Jouzel, Carsten Reinhardt, Alexander von Schwerin and Didier Torny have been especially helpful and stimulating.

We are also grateful to David Cantor and the anonymous referees for their insightful comments. Finally, we wish to thank David Cantor and Ruth Ireland for the smooth running of the project. Their professionalism made it a real pleasure to edit this book.

LIST OF FIGURES AND TABLES

LIST OF CONTRIBUTORS

Soraya Boudia is Historian of Science and Professor of Sociology at the University of Paris-Est Marne-la-Vallée. Her scholarly work focuses on the transnational government of technological and health environmental risks. She is the author of *Marie Curie et son laboratoire. Science et industrie de la radioactivité en France* (Paris: Editions des archives contemporaines, 2001). Among the volumes she has co-edited are 'Risk and Risk Society in Historical Perspective' (*History and Technology*, 23:4 (2007), pp. 317–31) and 'Transnational History of Science' (*British Journal of History of Science*, 45, Special Issue 3 (September 2012), pp. 319–36). She is currently writing two forthcoming books, the first one on scientific expertise and nuclear risks and the second one on expert knowledge in the governability of democracies.

Emmanuel Henry is Associate Professor of Political Science at the University of Strasbourg. His work analyses the public policies in the field of public health and occupational health in France and in Europe. He is the author of *Amiante: un scandale improbable. Sociologie d'un problème public* (Rennes: Presses Universitaires de Rennes, 2007) and the co-editor, with Claude Gilbert, of *Comment se construisent les problèmes de santé publique* (Paris: La Découverte, 2009). He recently published 'A New Environmental Turn? How the Environment Came to the Rescue of Occupational Health: Asbestos in France, 1970–1995', in C. Sellers and J. Melling (eds), *Dangerous Trade: Industrial Hazards across a Globalizing World* (Philadelphia, PA: Temple University Press, 2011).

Nathalie Jas is a Senior Researcher at the French National Institute for Agricultural Research (INRA). An historian and a STS scholar, her scholarly work analyses the intensification of agriculture and its social, environmental and health effects. She is the author of *Au carrefour de la chimie et de l'agriculture : les sciences agronomiques en France et en Allemagne 1840–1914* (Paris: Editions des archives contemporaines, 2001). She is the French translator of Ludwik Fleck's pioneer work *Genesis and Development of a Scientific Fact – Genèse et développement d'un fait scientifique* (Paris: Les Belles Lettres, 2005). Among the volumes she has co-edited are 'Risk and Risk Society in Historical Perspective' (*History and Technology*, 2007) and *Penser avec Fleck–Investigating a Life Studying Life*

Sciences (Zurich: Collegium Helveticum, 2009). She is currently completing her forthcoming book, a history of the government of sanitary risks posed by the agricultural pesticides in twentieth-century France.

Michelle Murphy is a historian and feminist techno-science studies scholar whose work explores conjunctures of sex, reproduction, race, environment, economic practices and techno-science in the late twentieth century and contemporary times. She is the author of *Sick Building Syndrome and the Problem of Uncertainty: Environmental Politics, Technoscience, and Women Workers* (Durham, NC: Duke University, 2006), *Seizing the Means of Reproduction: Entanglements of Feminism, Health, and Technoscience* (Durham, NC: Duke University, 2012) and the co-editor of the 2004 issue of *Osiris*, 'Landscapes of Exposure: Environment and Health in Historical Perspective'. She is Associate Professor at the University of Toronto.

Christopher Sellers is Professor of History at Stony Brook University. He holds a PhD in American Studies as well as an MD and is the author of *Hazards of the Job: From Industrial Disease to Environmental Health Science* (Chapel Hill, NC: University of North Carolina Press, 1997) and *Crabgrass Crucible; Suburban Nature and the Rise of Environmentalism in Twentieth-Century America* (Chapel Hill, NC: University of North Carolina Press, 2012). Among the volumes he has co-edited are the 2004 issue of *Osiris*, 'Landscapes of Exposure: Environment and Health in Historical Perspective', and *Dangerous Trade: Histories of Industrial Hazard across a Globalizing World* (Philadelphia, PA: Temple University Press, 2011). His forthcoming book is on the relationship between suburbanizing, race and environmentalism around Atlanta, and he has also begun work on a new project, a comparative and transnational history of industrial hazards in the US and Mexico.

Sezin Topçu is Researcher at CNRS (French National Research Centre) and a member of the Centre for the Study of Social Movements (CEMS-Ecole des hautes études en sciences sociales). Her research focuses on technological risk issues, environmental controversies, social movements and scientific governance. In her PhD thesis, entitled 'L'agir contestataire à l'épreuve de l'atome. Critique et gouvernement de la critique dans l'histoire de l'énergie nucléaire en France (1968–2008)' (Paris: EHESS, 2010), she analysed the transformations of criticism and its management in the French nuclear energy domain since the late 1960s.

Didier Torny is Senior Researcher at the French National Research Institute for Agricultural Research (INRA). He holds a PhD in Sociology and is the co-author of *Les Sombres précurseurs. Une sociologie pragmatique du risque* (Paris, Editions de l'EHESS, 1999). His work examines private and public normative action relating to health, i.e. norms which attempt to impose human or animal health as a legitimate objective for regulation. His research is based on a comparative pragmatic approach (nutrition, food safety, animal health, safety of health products), analysing the argumentary and critical mechanisms at work in definitions and written texts on these norms, from the simplest of recommendations through to international agreements. He is especially interested in the question of emerging diseases, both as a category of general action and as a series of pertinent examples through which to get to grips with the mechanisms used in collective risk management.

INTRODUCTION
SCIENCE AND POLITICS IN A TOXIC WORLD

Soraya Boudia and Nathalie Jas

In April 2004, as discussions reached a conclusion on the new European Community regulation on chemicals and their safe use (EC 1907/2006), dealing with the Registration, Evaluation, Authorization and Restriction of Chemical substances (REACH), the WWF publicized the results of blood tests on thirty-nine European Members of Parliament (MPs) who were not exposed to any particular chemical pollutants. No fewer than 76 persistent toxic chemical substances were found in their blood, out of the 101 for which they had been tested. On average, each MP carried a cocktail of forty-one toxic products composed of substances that were persistent (not biodegradable) and bio-accumulative (that accumulate in the body). Thirteen chemical residues (phtalates, perfluorinated components) were systematically found in all the MPs' blood samples.[1]

Since the early 1990s, revealing the systematic presence of toxic chemical substances in each individual's body and measuring the 'toxic body burden' has become one of the main means of action for activist organizations like the WWF. But the toxic body burden is more than that; it has also become a research subject for public health specialists, doctors, toxicologists and biologists increasingly interested in the effects of substances present almost permanently in human bodies. Surveys highlighting the increasing contamination of human bodies are also undertaken by government institutions such as the US Center for Disease Control (CDC), which has set up programmes that regularly measure and publish results confirming the presence of numerous chemical contaminants in human bodies.[2] With the toxic body burden, regulatory authorities, especially at international level, are faced with new challenges. It is compelling them to set up bio-monitoring systems that still have to be defined from A to Z – identification of the substances to be tested, the methods, etc. – and that imply many difficulties, notably with regard to the use of the results.[3]

The emergence and increasingly frequent reference to the toxic body burden are signs, among many others, that the massive presence of chemical substances in the environment, to the point of being in each human body, has shaped what is now called a 'toxic world'. This world is characterized by an uncontrolled and uncontrollable proliferation of chemical substances and physical agents, an accumulation of exponential quantities of toxicants in the environment, and an ever-growing range of effects on health: cancers, various allergies, disruptions of hormonal functions, trans-generational effects, auto-immune illnesses, developmental disorders, sick building syndrome, multiple chemical sensitivities, and so forth.[4] We see the advent of this toxic world as the result of two dynamics. First, the boom in industrial activity and the development of techno-scientific activities over the course of the twentieth century led to the introduction of thousands of chemical substances into the environment. These substances have continually extended the range of potentially dangerous health and environmental hazards. Second, since the nineteenth century, the regulation of toxic substances has not attained its stated objectives, that is, to control the dissemination of toxic substances and prevent their harmful effects on health and the environment. In this context, the development of the Global Harmonized System by the United Nations, the implementation of REACH in Europe from 1 January 2007,[5] and the amendment currently underway to the US Toxic Substances Control Act (TSCA), first passed in 1976,[6] seem to be part of a new attempt to develop a management of toxicants that is effective, and to solve many of the problems that these substances are causing in today's world.

The aim of this volume is twofold: to shed new light on the systems of regulation of toxic substances as they were developed throughout the twentieth century; and to analyse the new directions these systems seem to be taking at the beginning of the twenty-first century in order to cope with the changing nature of the problems generated by toxicants. It focuses on what can only be called a failure, in so far as these regulations have not prevented an unprecedented growth of chemical contaminants and the resulting set of problems that are now difficult to control. In particular, the papers in this volume examine the logics underpinning the development of these regulatory regimes through an analysis of the tools that they have developed and implemented. Together, they show that the failure of these tools is currently resulting in an historical shift in the politics of toxicant regulation, which until now has not been characterized. After World War II, the politics of toxicants was based on a desire to master contaminations and their side effects. Until recently, the stated objective was to protect populations or to keep pathogenic effects at acceptable levels. The prevention of contaminations is now being replaced by the requirement to regulate a continuously expanding number of toxicants in a world which is already toxic, and to ensure that populations accept and learn to live and cope with generalized contamination.

In the following section, we would like to examine the core issue of this book: the paradox of the development of increasingly complex systems for regulating toxicants, and simultaneously of an increasingly, irremediably toxic world requiring new regulatory strategies.

Regulating Toxicants: An Old Issue

Current regimes of regulation of toxic substances and the sanitary and environmental risks that they spawn are the outcome of events dating back to long before 1945. The different modalities of management that were experimented with and developed in certain periods have gradually been superimposed on and entangled with one another, leading to a variety of approaches to get rid of toxicants and their deleterious effects on health and the environment – or at least to control and contain them, make them invisible, or find a way of living with them.

Since the nineteenth century, rapid industrial change has profoundly altered the environment, bring with it chemical pollution, technical accidents and the poisoning of the bodies of workers, residents and consumers alike. These multiple effects have not been overlooked. They have triggered numerous debates and types of management: expert commissions, especially within academia, court cases, insurance policies, compensation, improvement of technical systems to limit emissions or their effects, development of sets of regulations to frame use, and new administrations dedicated to the management of potentially dangerous substances.[7]

Additionally, a management of sanitary and environmental dangers was set up, based on the idea of controlling flows of contaminants and their effects by limiting exposure levels. As a result, regulatory systems have been designed less to prohibit the use of substances recognized as dangerous but often considered as indispensable, than to confine them. The first types of confinement aimed at were geographical and social. For example, laws on classified sites (supervision of factories presenting risks) – the oldest of which were French and date back to the very beginning of the nineteenth century – were intended to group together dangerous and polluting activities far away from wealthy neighbourhoods, in poor areas which also provided labour. These territories and their populations thus constituted environments and exposed bodies that were left to absorb contaminants. However, these confinement strategies were only partially effective, for contaminating flows could never be controlled completely. Factories' discharge of toxic substances into the environment, or the presence of such substances in foodstuffs or in manufactured products, spread their effects by affecting increasingly large populations, including the wealthiest.[8] From the 1860s, in addition to confinement, the challenge was therefore also to control the 'overflowing' of these chemical industries. This type of control was implemented through policies which aimed to reduce their effects.

Although laws in this respect had been passed from the early nineteenth century, from 1870 the implementation of regulatory systems accelerated. This corresponded to a period during which, in general, the state was expanding its ambit and simultaneously changing its methods, notably by developing new administrations in which technical and scientific expertise played an essential part. The last third of the nineteenth century and the early twentieth century was thus a period in which the foundations were laid for many national regulatory systems, notably with regard to foodstuffs, medicines, professional medicine, toxic substances and industrial pollution.[9] The establishment of these systems coincided with the development of new scientific disciplines such as industrial hygiene and toxicology, as well as new professions such as jurists, expert chemists and inspectors working within these institutions. But although these regulatory systems became stronger during the inter-war period, they failed to prevent sanitary scandals resulting from the development of certain sectors of activity: pollution through industrial accidents, and collective poisoning through pesticides or medicines.[10]

In the wake of World War II the scale of problems posed by toxicants changed radically. As a result of the development of the nuclear industry and synthetic chemicals, the dangerous substances with which humanity had to deal proliferated. The world witnessed an unprecedented increase in the quantities of chemical substances put into circulation and onto the market, and some of those substances started to be found in the atmosphere, the soil and water. Most of them had not been evaluated or regulated in any way whatsoever. The regulatory systems under reconstruction or creation had to face a continuous flow of substances that were already in circulation, as well as the new substances that were regularly put on the market. Pressure on the regulatory authorities intensified, especially since, as Soraya Boudia's paper in this volume shows, with the growth of the nuclear industry they were gradually confronted with growing concerns by some scientists who feared the effects of the proliferation of dangerous substances on the ecosystem as a whole. This paper highlights the decisive role of nuclear power in the emergence of the idea of gradual poisoning of the planet by the chemical contaminants that had to be contained.[11]

Probably the most noteworthy feature of changes in the regulation of toxic chemical substances was the beginning of its internationalization. During the 1950s the new international institutions – UN organizations like the FAO (Food and Agriculture Organization), the WHO (World Health Organization), the ILO (International Labour Association) and the AIEA (International Atomic Energy Agency), as well as organizations resulting from European construction – set up a series of committees of 'international experts'. These institutions, along with professional societies, some of which had close ties with industry, for example the International Union of Pure and Applied Chemistry (IUPAC), and academic societies, for example the Union for International Cancer Control (UICC), as well as international organizations of producers or users of chemical substances, for example the Permanent International Committee on Canned Foods, were behind the initiative of many scientific events or meetings between experts on the problem of the effects of toxic substances. Christopher Sellers's and Nathalie Jas's papers in this volume discuss the early development, in the 1950s and 1960s, of an international order for regulating toxic chemical substances. They pave the way for research on this internationalization which until now has received little attention in the literature. The comparison of these two contributions shows that this internationalization was promoted by a handful of scientists, considered at the time to be 'international experts', who circulated between sectors – and between national and international levels – and thus contributed to the circulation of ways of thinking and of tools for regulating toxicants.

A variety of tools for managing toxicants and their effects was developed in the 1950s in the national and international regulatory systems – some of which had already been experimented with before the war. These included: the creation of mechanisms to frame the use of toxic substances, for example through labelling or specific protection measures; the 'specification' of substances, indicating, for example, their precise characteristics and the levels of impurity authorized, or the methods authorized for the evaluation of their toxicity; the production of lists of substances that met specific criteria and could be authorized under certain conditions; the classification of substances by identifying the most dangerous ones that should be examined and regulated first; and the production of indicators and limit values to guarantee that exposure would not have irremediable effects on health.

Of all the tools structuring the regulation of toxicants in the latter half of the twentieth century, limit values played a very specific, crucial role, as this book demonstrates. By analysing their trajectory, the papers in this volume reveal the logics underpinning the regulation of chemical substances, as well as their political implications.

Regulation Through Thresholds

The idea of setting limit values for exposure was born long before the 1950s. This mode of action has a long history dating back to at least the latter third of the nineteenth century, when discussions were held on water pollution and occupational health.[12] Traces of such discussions can also be found elsewhere, in debates in the early twentieth century on insecticides on fruit.[13] Above all, setting the maximum quantity that could be administered to a patient became an important issue with regard to poisons such as arsenic or mercury, which were then used as medicines. In the inter-war years norms for administering this type of substance existed already. Yet the logic of setting a threshold, which became characteristic of the systems of regulation of the latter half of the twentieth century, started to be concretized only from the beginning of the 1940s in the US in the field of industrial hygiene. At the time, the American Conference of Governmental Industrial Hygienists (ACGIH) started to produce Maximum Allowable Concentration values (MACs) for toxic substances present in work environments.[14] The concept of the MAC has since been highly successful. It has been taken up, revised and adapted in multiple ways, as is well illustrated by the contributions of Christopher Sellers and Emmanuel Henry to this volume.

MACs were based on the then fundamental dogma of toxicology, that 'the dose makes the poison' and that it was therefore possible to determine for each toxin a threshold under which no harmful effect could be found. Along with their derivatives, MACs served first to solve the problem stemming from the fact that substances could be toxic in large doses but not, or less so, in lower doses. They made the use of substances possible, without causing – at least theoretically – unacceptable sanitary effects in certain contexts such as factories, where exposure was not isolated, but regularly repeated.

The principle of the MAC was adopted from the mid-1950s in other fields, where it was used to produce tools for limiting the negative effects of the rapid deployment of the Chemical Age. This appropriation was swift and particularly successful in the sensitive food additive and contaminant field, as Jas's paper shows in this volume. In the period from the mid-1950s to the early 1960s, a group of international experts developed the Acceptable Daily Intake (ADI). This was an indicator that served to set norms, called 'tolerances', for food additives and contaminants, and guaranteed that the quantities present in foods were not dangerous. The production of limit values then spread to other fields, especially to assess the contamination of the air, water and soil. In the environmental field, these limit values were gradually formalized and, in the 1960s and 1970s, produced tools such as Maximum Concentration Limits (MCLs) and Toxicological Reference Values (TRVs).

The capacity of limit values to guarantee the innocuousness of toxic substances relied on the fact that, for each substance, there was a threshold under which no harmful effect could be found. Yet from the early 1950s, that is, precisely the time when a mode of government through thresholds was being developed, scientific studies started to show that it was impossible to determine exposure thresholds for carcinogenic substances. This shift is at the core of the analysis proposed by Boudia's paper. She explains how the idea generally accepted before World War II, of the existence of a threshold below which there was no risk of exposure to radioactivity, or only a negligible risk, was seriously challenged by the experts after World War II. What Boudia shows, most importantly, is that the significance of this result and its appropriation by regulatory authorities stemmed from the context of controversy and active protest against nuclear power during the Cold War, which compelled the regulatory authorities to publicly recognize what until then had been confined to expert circles. Apart from the nuclear case, all industries producing or using carcinogenic chemical substances were concerned by this problem. While a conservative logic argued for the banning of all substances recognized as possibly or potentially carcinogenic, that would have meant banning many substances that were indispensable to certain industrial sectors.

Hence, from an economic point of view, a ban was generally impossible to envisage. National and international regulatory bodies therefore rapidly acknowledged, albeit in different ways depending on the context, that carcinogens could be authorized in proportions set by limit values. A well-known example of this was the Delaney Clause passed in 1958 in the USA. This law banned the presence in food of substances found to induce cancer in humans or, after tests, found to induce cancer in animals. Yet it was rarely applied and was amended many times (especially with regard to pesticide residues). Since the 1950s, the problem posed by carcinogens and their non-threshold features has become a generalized issue, constantly generating debate and arrangements to contain the danger posed by these toxic substances and their widespread use. The papers in this volume reveal the wide variety of these arrangements. Thus, Jas tells us how, from the early 1960s, the young Joint Expert Committee on Food Additives created by the FAO and the WHO went to great lengths to recognize the impossibility, for socio-economic reasons, of totally excluding carcinogenic substances from foods. With regard to carcinogens in the work environment, several states endorsed the necessity to accept the presence of non-threshold carcinogens in work environments. In so doing, they recognized that for these substances, exposure limit values were based not on scientific data but on socio-political considerations. Henry underlines the fact that the strategy chosen by the French state was very different. He describes how, for a long time, it refused to officially set limit values for carcinogens in occupational contexts, thus avoiding public acknowledgement that certain workers were wittingly exposed to this type of substance.

The varied cases analysed in this volume show that the regulation of toxic chemical substances was based on numerous compromises to suit the needs of industry, and that it clearly did not hinder the development of the chemical industries, nor that of the many industries using chemical products. This volume also demonstrates how limit values enable us to understand the multiple ways in which these compromises were reached. The main characteristic not only of the limit values tool but also of all the management tools found in toxicant regulations implemented after 1945 – such as lists of substances authorized or banned, lists of substances that were of particular concern and needed to be studied as a priority, and specifications of substances – are their remarkable malleability and thus their adaptability to widely diverse situations. Studying the case of the Chemical Valley, which stretches along the St Clair River in Ontario and some great lakes, Michelle Murphy's paper shows that these tools not only impose a substance-by-substance, activity-by-activity reasoning concerned with short periods; they also have the advantage for industry and regulatory bodies of side-lining and rendering invisible the sensitive problems posed by phenomena of accumulation of toxic substances and of interaction between them.

This malleability enabled the systems that applied these tools to shape them according to their own requirements and constraints. From this point of view, the two papers in this volume that discuss the production and use of limit values in occupational contexts are illuminating. Limit values became a particularly successful tool from the 1950s for controlling workers' exposure. As Sellers shows, each of the elements comprising them could be arranged in multiple ways, and incorporate the workers' social representations and the scientific practices of the countries or regions in which they were produced. Thus, experts from the US based their levels more on epidemiology and frank clinical disease; those from the Soviet Union, more on that nation's unique brand of animal toxicology; with the Europeans somewhere in between. Limit values can then be integrated into national systems that have different characteristics as regards care for occupational diseases, and into the different relations that the state has with the protection of a common good such as health.

The malleability of regulatory tools has an essential virtue. Irrespective of the geographic location or the period concerned, it usually makes it possible to unite two imperatives of a different, often contradictory nature: the protection of public health and, less visible but extremely restrictive, and usually a priority, the necessity not to place particularly strong constraints on the industries concerned. The regulation of toxic substances has thus been designed above all to provide solutions to the problems experienced by industry, due to the harmful effects on human health and on the environment of toxic substances considered as necessary. Even though they had the secondary benefit, in the latter half of the twentieth century in Western countries, of contributing to the reduction of the deleterious effects of chemical substances, especially the most visible acute cases of poisoning, these regulations have above all allowed for the most cost-effective management of dangerous substances in the workplace, in consumer products, especially food and medicines, and in the environment: water, air and soil. Their existence has thus helped to naturalize the presence of an increasing number of toxicants in everyday consumer products and in the environment.

In a world where the economy has a central role, the capacity of regulatory systems to preserve or even to facilitate certain types of economic and social development is crucial to understand their success and the importance they have taken on. The fact that, ultimately, the protection of health and the environment are not their main function has probably been the cause of recurrent crises of the same nature in the long term, along with the maintenance of other forms of risk management, notably court cases, and the development of forms of radical protest aimed at drawing attention to these dangers that the regulatory systems end up concealing in certain cases.

Transformation and Inertia of Regulatory Systems

Under severe criticism and targeted by environmentalists of all kinds, the systems regulating chemical substances in the post-World War II years, like those of all techno-scientific and industrial activities, experienced their first major crisis at the end of the 1960s.[15] First in the USA and then in certain European countries, this crisis led to substantial changes in regulations and institutions and, most significantly, to the creation of agencies dedicated to the management of environmental problems.[16] During the same period several transnational initiatives were also taken with regard to the issues of chemical substances, the control of their circulation, and the management of the wide range of problems they involved. For instance, a unit devoted to carcinogens, responsible for evaluating the carcinogenicity of chemical substances, was included in the International Agency for Research on Cancer (IARC), under the World Health Organization, from the Agency's inception in 1967.[17] The creation of the United Nations Environment Programme (UNEP) in 1972 set a new framework for the treatment of health and environmental problems. Under its aegis, the UN system of control of toxic chemical substances was entirely reviewed in the second half of the 1970s. This led, in particular, to the creation of the International Programme on Chemical Safety (IPCS), entrusted with the evaluation of the toxicity of chemical substances. Europe was also part of this movement. It passed specific laws on chemical substances and gradually engaged in policymaking based on directives, which were binding on member states.[18] While this period saw the proliferation of regulatory bodies, industrial actors nevertheless retained a central role in the definition of public policies on toxicants. Having adopted a lower pubic profile, they adapted to the new political and institutional configurations by actively engaging in lobbying activities and multiplying ties and interaction with regulators.

The crisis in the management of dangerous substances at the end of the 1960s showed that the measures in force were insufficient and unable to withstand criticism or to limit the dangerous effects that were revealed, inter alia, by studies on cancer and the environment. The activist movements that pointed out these inadequacies were backed up by scientists who called for changes in the evaluation and regulation of chemical substances, on the basis of new scientific knowledge. Thus, to the study of carcinogenesis was added that of eco-toxicology[19] and environmental mutagenesis,[20] which contributed to highlighting new effects of dangerous substances for living beings, and led to the elaboration of a new system of classification of dangerous substances. One of the characteristics of the new effects of toxic substances was the much broader spatial and temporal scale on which they operated. Pollution was no longer only local, it could affect the entire planet; and it concerned not only health but the whole ecosystem. Moreover, the consequences were not only immediate; they could last for generations. Precisely because they occurred on previously unknown scales, from the infinitely small to the infinitely big, the sanitary and environmental effects of toxicants confronted experts and institutions with an unusual set of problems.

One of the thorniest issues at the time concerned the effects of exposure to low doses of pollutants. As Boudia's paper demonstrates, the nuclear industry was the matrix within which the question of effects of exposure to low doses emerged, and scientific and regulatory concepts to study and manage it were defined and implemented. The issue of low dose effects involved scientific problems to which researchers and doctors had trouble finding satisfactory solutions. This posed an important political problem for policymakers responsible for the management of toxic substances. The formulation of the problem of low doses amounted to saying that innovations could have harmful sanitary and environmental effects, not only in exceptional situations such as accidents but also in 'ordinary' situations, in a context of normal and controlled functioning of an activity. This was an irreducible criticism targeting various scientific and industrial fields. Since for political and industrial decision-makers and the trade unions it was out of the question to even envisage ceasing those activities in which the problem of low doses was an issue, tools had to be found that could justify their pursuit and thus provide answers to the public criticism levelled at them.

Boudia's paper shows that recognition of the potential problem of exposure to low doses of pollutants generated a contradiction in the practices of regulatory systems: whereas it amounted to acknowledging that there was no threshold below which a substance could be considered harmless, the setting of limit values nonetheless remained at the centre of regulatory systems. To overcome this contradiction, the procedures for setting limit values were from then on claimed to be designed to guarantee not the absolute innocuousness of substances in certain conditions of use, but instead 'tolerable', socially 'acceptable' levels of risk. Hence, it was recognized that exposure norms were not only the fruit of scientific decisions but also that they incorporated economic and political considerations.[21] The institutional changes in the 1970s and 1980s fully integrated this new dimension, expressed in the separation of the 'assessment' of substances from their 'management', formalized by the NRC Red Paper on risk assessment.[22] This separation between a scientific evaluation phase and a phase concerning decision-making on the policies to implement is at the heart of current regulations on toxic substances. It first appeared in the US, and then in the 1980s and 1990s gradually spread to the international institutions, which required the countries that had not already adopted it to do so.[23] The importance of this logic underpinning the functioning of toxicant regulation systems, and of the role of international institutions in its implementation, certainly warrants further study.

These shifts in the logics underpinning the management of toxic substances took place in the framework of unprecedented growth of regulatory authorities. The result was regulatory systems for chemical substances that were complex and multi-level, and in which mechanisms of inter-regulation were set up. The two last papers of this volume illuminate how these mechanisms work, between collaboration and competition. Sezin Topçu describes a dynamic of collaboration between different regulatory levels followed by the ETHOS project that she studied. This project, aimed at developing a new approach to rehabilitation of highly contaminated areas in the Chernobyl region, rapidly received the support of major institutions, both national (French and Belarus governments) and supranational (European Commission, World Bank). Didier Torny's case study is exemplary of phenomena of competition between different regulatory levels and between international institutions that have also emerged. Within a period of three years, and whereas scientific knowledge on this substance had not fundamentally progressed, three different Maximum Residue Limits (MRLs) were produced for the pesticide chlordecone in the French Antilles. Each of these MRLs were primarily a response to different – and conflicting – political and social demands, and thus depended not only on the context but also on the preoccupations of the institutions that produced them – the French state or Europe.

By the early 1990s contaminants had become a lasting and ever-expanding problem at both national and international level, despite the existence of a 'more precautionary science' (at least since the 1950s, as Sellers shows), and despite decades of multiple scientific and political initiatives aimed at dealing with the problem of toxic substances. A renewal of the measures and tools for managing flows of toxic substances and their effects on health and the environment was once again called for. The fact that new substances with unknown properties were put on the market, along with the acceleration of the transfer towards the global South of some chemicals industries – often the most polluting ones – and the increasing industrialization of agriculture and food, were all parameters that widened the circulation of chemical substances and consequently the range of their most pathogenic effects.[24]

A series of intricate problems posed by these substances were increasingly discussed and were brought to the fore in public debates. Scientific studies showing an increase in the number of cancers imputable to environmental exposure proliferated, as did the warnings by experts in environmental health. These experts associated the development of new pathologies with the generalization of the presence of contaminants in the environment.[25] Environmentalist organizations and other NGOs drew on this expertise to renew their criticism of toxic substance management. One of their main criticisms of the regulatory systems was the recurrent incapacity to deal with the tens of thousands of substances that had been put on the market since World War II. When the White Paper that launched the process of reforming the regulation of chemicals within the EU was published, the European Inventory of Existing Commercial Chemical Substances (EINECS) counted over 100,000 chemical substances in circulation in the Union, most of which had not been evaluated at all.[26] The reform of the Toxic Substances Control Act (TSCA) of 1976 currently underway in the USA is based on the finding that there are over 80,000 substances in circulation in that country on which there is little or no toxicological data. In 1976 the EPA, in charge of the implementation of the TSCA, authorized, without any evaluation of their toxicity, 62,000 substances that were to remain on the market. Since then, of the 22,000 new substances that have been put into circulation in the USA, about 200 have been evaluated within the framework of this system, and only five have been regulated.[27]

All of this criticism again prompted national and international institutions to review their policies for managing contaminants and the long-term impact of those policies. This renewal has been clearly apparent for the past ten years. We have been witnessing the gradual elaboration of a transnational system of regulation of chemical products, with a process of consensus-building between different countries and the different actors concerned (public authorities, industry, NGOs). In 1995, several international institutions (UNEP, ILO, FAO, WHO, OECD, UNIDO)[28] set up the Inter-Organization Programme for the Sound Management of Chemicals (IOMC), in which the United Nations Development Programme (UNDP) and the World Bank have observer status. These institutions were thus responding to the recommendations adopted at the 1992 World Earth Summit in Rio, concerning the necessity for international cooperation to improve the protection of health and the environment against chemical substances. The main objective of this programme was to promote coordination for the harmonization of systems of classification of chemical products, and to develop a single system at international level. Seven years later, the work under this programme led to the first version of a generalized, harmonized system, the Globally Harmonized System of Classification and Labelling of Chemicals (GHS). In parallel, from 1998, faced with the constant failure of policies on the pathogenic effects of chemical substances, the European Community embarked on similar work to classify chemical products. Nine years later, after much lobbying and many public debates it ruled that the 'Registration, Evaluation and Authorization for Chemicals' (REACH) regulation had to be applied by all European countries. Since 2008, it has been mandatory for firms to carry out studies that document the toxicity of the substances they use or produce, and to register them with the European Chemicals Agency. The study of the preparation and implementation of REACH reveals just how influential industries producing or using chemical substances remain in the definition of regulatory procedures and policies. Following the publication of the White Book in 1998, which opened the process leading towards REACH, industrial actors lobbied extensively for the project to fail. From 2003, when they realized they would not be able to prevent the new regulation from being passed, they actively sought to obtain favourable amendments to the text that was being drafted.[29] They also prepared for the new regulation by transforming their production practices, substituting certain substances with others, or relocating production to countries of the South. Moreover, the trade unions of the chemical industries were very well organized and developed new lobbying systems in order to influence the implementation of the new regulation. They were active in the production of technical expertise to be taken into account for the classification of chemical substances according to the level of danger they represented, which in turn determined the level of evaluation required, as well as for the production of exposure limits.

One of the objectives of the reform of regulatory systems currently under way is thus to have an evaluation of the toxicity of all the substances already in use, to set up systems for evaluating all new substances, and thus to rule on all the chemical substances in circulation, on an ad hoc basis. This project has enthused many researchers and has been endorsed by most NGOs, who see it as a clear departure from the policies on chemical substances implemented hitherto. But on closer inspection we see that the tools developed to meet these objectives, which made the REACH's reputation, do not seem in any way innovative and are largely inspired by those used previously. The aim is still to categorize the substances considered to be the most dangerous and which therefore have to be treated as a priority. This work of prioritization, carried out before any evaluation, is long and political. As past experience has shown, it drastically reduces the number of substances finally concerned by in-depth evaluations. Likewise, one of the REACH's leading measures – putting the onus on industries and not regulatory agencies to prove the absence of dangers of chemical substances – was presented as a major innovation, which has been taken up by countries reforming their toxic substances regulation.[30] Yet it has already been used experimentally for substances such as medicines and pesticides, for which authorizations are required for putting new products on the market. For these two categories of chemical substances, the obligation that firms have to carry out toxicity evaluations has not prevented large quantities of substances which posed and regularly still pose sanitary and environmental problems, some of which are very serious, from being put into circulation. Furthermore, the reforms currently underway aim at producing regulatory systems with a global reach, underpinned by the desire to articulate and harmonize the different existing regulations. Aside from certain regulatory bodies' desire to extend their influence, such ambition also reflects the multitude of regulatory systems in place. These different systems operate at difference levels: regional, national, international; they are specific to certain domains: the workplace, the environment, food; or to certain types of substances: pesticides, medicines, persistent organic pollutants, and their legal status and the level of constraint they impose vary widely. The tensions and contradictions between the different regulatory systems become particularly apparent at the local level where precise issues are addressed, as Michelle Murphy's paper highlights. The intertwining of different regulations, stemming from a specific history, renders the regulation of chemical substances indecipherable to many actors concerned by their effects. This has at least two consequences: first, it helps the minimization of constraints on industrial actors, and, in turn, of the level of protection of populations; second, it contributes to making the health and environmental problems raised by toxicants institutionally invisible.

For the past six decades at least, the continuous flow into the environment of a huge number of chemical substances with various properties, which has been permitted by regulation, is generating increasingly insoluble problems. Yet, as ambitious as they may seem, current reforms to the systems regulating toxicants reflect many forms of inertia.[31] The critics are right, and what they are demanding is above all a significant reduction in the number of substances effectively in circulation, as well as the quantities used.[32]

Living in a Toxic World

Current reforms to the systems regulating toxicants are encountering many limitations. The complex materiality of the toxic world, as well illustrated in Murphy's paper, means that the challenge is not only to evaluate and rapidly take measures to ban or regulate diverse uses of thousands of chemical substances in order to limit their harmful effects. This volume makes the point that faced with the impossibility of rehabilitating contaminated sites, controlling contaminations and preventing or repairing their sanitary effects, governing bodies have to increasingly manage the lives of people within spaces subjected to various forms of exposure to toxic substances. The two last papers in this volume show that certain highly contaminated sites thus serve as laboratories to experiment with new ways of managing these contaminants and their effects on health and the environment. New management methods are tested to organize life in permanently contaminated territories, especially in areas which have experienced major accidents or an irreversible accumulation of toxic chemical substances that have penetrated the ground and contaminated the water. Torny and Topçu's papers discuss such experiments, opening a new field of investigation. The case, developed by Torny, entitled JAFA, is a state-funded community health programme in the French Antilles, contaminated by large quantities of a pesticide in the persistent organic pollutant category. The second experiment, the European programme Ethos, analysed by Topçu, concerns the Chernobyl area after the 1986 accident. This project has combined economic development and health protection. Murphy's paper shows that some of these tools are used not only in post-disaster life but also in the 'normal management' of areas contaminated through a concentration of industrial activities, such as those of the petro-chemical industry.

The analysis of these experiments reveals a transformation underway in the logics underpinning toxic substance management. The first characteristic of this transformation is that it is now necessary to accept and to live with contaminants that are already there, since their confinement and control, and the control of many of their effects, has failed. The second characteristic is that their management is not only the responsibility of public administrations and institutions; it also involves the populations affected by these contaminations. This so-called participative management, inspired by an unprecedented growth in recourse to participation, is based on the invention of new regulatory tools that incorporate existing ones such as limit values, but concentrate mainly on the production of rules of conduct. The third characteristic of this transformation is the focus on the individual and on an individualization of the modalities of management accompanied by discourse on everyone's personal responsibility.

These new modes of management based on concrete measures and rules of conduct enable everyone to live with as little discomfort as possible, both now and in the long run, in a world lastingly contaminated to varying degrees. These transformations are accompanied by a transfer of responsibilities from the international organizations and the states and their regulatory agencies towards every individual who is attached to a new way of thinking and acting. Individuals are enjoined to take charge of their own management of contaminants and the risks they involve. Torny shows how the JAFA focuses on the consumers of home-grown products and small-scale farming, establishing their level of risk and advising people about safer agricultural and consumption practices. Topçu explains how, with the Ethos project at Chernobyl, everyone is asked to be 'autonomous' and 'participatory'. These experiments, and others, that would require investigation, seek to make every individual an active player in their own life. To that end, they multiply the guidelines on how to choose one's food and how to behave in a permanently contaminated area. The inhabitants are thus given information to be able to limit as much as possible their exposure and thereby the effects on their health.

These experiences and others show that the art of governing populations, territories and problems increasingly consists of influencing the behaviours of those affected in various ways by toxic substances. The 'conduct of conducts' currently being established in areas that are lastingly contaminated corresponds to the neo-liberal governmentality – as Foucault[33] and governmentality studies[34] have defined it – at work in other sectors and particularly in the fields of health and of crisis and disaster management.

This type of management can be compared to more wide-ranging changes in the field of health, with the growth of a government of the self in which everyone must consider their body as a capital that has to yield returns.[35] Yet, as a tool used to manage a polluted environment and to limit the effects of contamination, making one's individual capital yield a profit has a particular meaning: it makes everyone responsible for their own 'toxic body'. Above all, this entails making everyone their own risk manager, defining and implementing solutions that should enable them to limit exposure to contaminants since it is not possible to avoid them. This situation is not, however, free of paradoxes, or even cynicism. The recognition of each individual's own responsibility can be presented as a liberating solution, empowering individuals to deal with a situation for which they are not responsible and on which they have only a very limited influence. Yet, with the toxic world, this interpretation of the implementation of neo-liberal governmentality is confronted with the fact that, in many cases, not only are those who suffer from the effects of toxicants not responsible for their production, they also have no alternative but to carry on living in contaminated areas and in conditions that are potentially or effectively harmful to their health. In these circumstances, what does it mean to be 'in control of one's life', for populations, often the poorest – like those in the French Antilles that Torny studies, or those in Belarus that Topçu discusses – with no possibility of leaving the dangerous territory in which they live, or rehabilitating it?

This new management of chemical contamination which, unable to control and contain, focuses on the organization of 'living with', can also be likened to what has been set up since the 1990s in the international management of crises and catastrophes. In this respect, the concept of 'resilience',[36] after that of 'vulnerability',[37] occupies an increasingly important place. Introduced by the United Nations after it declared the 1990s the International Decade for Natural Disaster Reduction Effort,[38] the concept of resilience has been applied in the construction of a new discourse calling for a new approach or a new culture in the mode of managing disasters, whether they are natural or the result of industrial or techno-scientific activities. Expressions such as 'sustainable and resilient communities' or 'binding community resilience' abound in reports and in the research literature. The implementation of resilience strategies bears witness to important shifts: it is a matter no longer of working to eliminate disasters but of recognizing their ineluctable nature and of endeavouring to reduce their impacts. This approach has been adopted in the plans to combat climate change, which take note of what global warming will lead to, and the fact that it is therefore necessary to find the means to reduce its negative effects.

Irrespective of the enthusiasm of those who defend it, the logic of 'living with' attests to a long-term failure of all the strategies that have been deployed to control the dangers presented by chemical substances. Rather than seeking to prevent and reduce them, it is now a matter of considering them as part of our reality. The role of humans is thus to reduce the disastrous consequences, to limit the negative effects as much as possible. Some may see this as evidence that the reality is taken into account, and that society is building on past experiences. Yet, can we really equate an earthquake or a tsunami to a nuclear or major industrial accident, or to long-term pollution with serious effects? The gradient of differences between various types of event is the degree of human intervention, the fact that many problems that we have to manage are the result of our own activities.

The Papers

Before concluding this introduction, we would like to introduce each of the papers and look at how, individually and together, they open new perspectives on the transformations of toxicant issues during the twentieth century, and on what is now at stake in the regulation of chemicals and their deleterious effects.

The volume opens with Christopher Sellers's paper in which he discusses the beginnings of the MAC (Maximum Allowable Concentration), a highly symbolic tool for regulating toxicants. As we know, the main idea underpinning the MAC is that it is possible to determine a threshold under which workers' health would not be affected by regular, repeated exposure to toxic substances. Sellers highlights four important points concerning the construction of contemporary systems of regulation: the centrality of threshold-oriented thinking; the importance of the occupational health sector in developing this thinking and in its early implementation in real life situations; the fact that the 1950s and 1960s were a turning point; and the complex articulation between transnational dynamics and local logics at work in the elaboration of these systems. He starts by analysing a surprising international conference in Prague in the spring of 1959, right in the middle of the Cold War, when representatives of twenty-six countries came together to negotiate international standards for workers' exposure to the most toxic substances. By analysing in detail the different ways used to set MACs, which were discussed at this conference, Sellers shows that the specialists from the different countries shared the same logic of precaution. However, he highlights the fact that, although US occupational health scientists had been the first, in the early 1940s, to formalize and produce limit values for exposure in occupational settings, the Soviet Bloc and Eastern European countries had also developed their own approaches. Sellers thus shows that conceptions of workers' bodies, of exposure, of protection and of techniques for studying and measuring toxicity, which differed widely depending on the geographical area and political culture, resulted in modalities for producing MACs that differed to those developed in the US. It was these differences that resulted in multiple exposure norms for the same substances, which led to the Prague conference in the context of emerging international regulatory systems. The main architect of this and subsequent conferences on the same subject in the 1960s was French toxicologist René Truhaut.

Truhaut played a crucial part internationally, in organizing dialogue between doctors and scientists from different countries, with a view to standardizing the modalities for setting limit values. His role highlights the importance of a few key players who shaped international expertise on toxic chemical substances in the 1950s and 1960s. This is the focus of the second paper in this volume, by Nathalie Jas. She examines the construction of the first tools for regulating food additives and contaminants, an industry which – second only to the nuclear industry and along with occupational health – was one of the main loci for reflection on and development of tools for regulating toxicants in the latter half of the twentieth century. This construction took place within international scientific committees, notably those created by UN organizations such as the FAO and the WHO. In this paper Jas highlights the importance of looking at how some of the foundations of international systems for regulating toxicants, which developed significantly in the second half of the 1970s, were laid during the preceding two decades. She also highlights how certain young scientists from Western Europe tried to take advantage of the construction of these international organizations to build spaces of legitimacy for their disciplines, especially toxicology, while developing their own careers. Finally, the approach via the construction of regulatory tools shows the contradictory logics underpinning the early development of this international expertise. Within a few years, from the late 1950s to the early 1960s, the concern of these 'international experts' to guarantee a high level of protection for public health had to make way for economic priorities. The experts were convinced that it was impossible to set a threshold for carcinogens beneath which no effect could be found, and that it was therefore essential to ban these substances in foods in order to guarantee the protection of public health. Nevertheless they eventually produced limit values for carcinogenic substances. They therefore contributed to the creation of a dynamic in which it was constantly necessary to find the means to justify the contradiction between the claim of protecting public health, and approval of the use of a multitude of dangerous substances. The risk technology that was gradually deployed throughout the 1980s was developed precisely to try to resolve that contradiction.

The third paper, by Soraya Boudia, looks at the growth during the 1970s and 1980s of these risk technologies which are now crucial in regulating the effects of toxicants. She highlights the centrality of the problem of the carcinogenic and mutagenic effects of low doses of radiation – which triggered many protests – in the emergence and structuring of these risk technologies. Boudia examines how the answers provided to this problem in the US and in international agencies in charge of regulating nuclear risks served as a matrix for formalizing reflection and devising a framework for analysing and managing a wide range of sanitary and environmental risks. To this end, she takes us through the complete genealogy of the 'Red Book' that formalized the risk-centred approach. This report, published in 1983 by the US National Research Council, is still the international reference in the regulation of toxicants today. Boudia's analysis shows how, as risk technologies matured, they came to be characterized by three main features. First, the regulation of toxicants has to be based not on the setting of a threshold corresponding to zero risk and guaranteeing the absence of all harmful effects, but on the setting of exposure norms corresponding to a 'socially acceptable' risk, in other words, a level of risk that generates no conflict or protest. Second, it is necessary to introduce a separation between two processes, the first scientific and the second political: 'risk assessment' determines the levels of danger according to the levels of exposure, and 'risk management' seeks to define a level of 'socially acceptable risk' – notably by setting the exposure norm that has to be retained. Third, risk assessment and risk management methodologies have to be generic, that is, not peculiar to one industry but applicable to all sanitary and environmental dangers.

Although risk technologies are crucial today in the management of toxicants and their effects, they were not taken up rapidly and unanimously everywhere. The international institutions and regulatory systems have constituted essential vehicles for the spread of these technologies, which sometimes require significant institutional and cognitive reconfigurations. This is the aspect that Emmanuel Henry's contribution illuminates by examining recent changes underway in France in the setting of exposure limit values for toxicants. Through the example of the occupational health sector, Henry shows that risk technologies were adopted in France only from the late 1990s, with the emphasis on the separation between assessment and management. This took place under the dual influence of the asbestos affair and pressure from the EU, which was seeking uniformity in the regulatory systems of the member states. Henry shows how these risk technologies, which make the expertise and decision-making processes more visible and explicit, conflicted with the French regulatory culture. For many years, negotiations on limit values had been based on debates between industry, the trade unions and representatives of the state. These three interests therefore strongly resisted this new system, which was not entirely able to impose itself, and the clash between old and new practices led to a local re-appropriation of risk technologies. Henry's paper highlights at least two phenomena. First, it shows the difficulty of transforming a regulatory system shaped locally over a long period. In doing so, it points out the fact that if risk technologies have circulated and have been so successful, this is probably partly owing to their generic nature. The fact that they are generic makes them sufficiently malleable to be re-appropriated locally and to contribute to reshaping regulatory systems, and able to bring their underlying philosophies and their modes of functioning closer together. Second, Henry's paper clearly shows that even though risk technologies have been discussed since the 1970s and have become so central in the governance of toxicants since the 1980s, the timing of their effective adoption has differed, depending on the country and transnational organization. Far from simply replacing the former regulatory systems, they tend to produce new systems through a process of hybridization with existing ones.

This entanglement of the timing and technologies of the governance of toxicants, stemming from different dynamics and currently at play in regulatory systems, is the core issue of the last three contributions to this book. The first paper is by Michelle Murphy. Studying the case of pollution in a petro-chemical area situated along the St Clair River in Ontario, Canada, and by the edge of a great lake, the paper's aim is twofold. It first describes and analyses what constitutes a chemically toxic environment, developed over more than a century of activity. The generalized contamination of the environment observed is not the result of an industrial disaster but the fruit of normal industrial activity, or at least one that is presented as such. Murphy rightfully stresses the significance of history in the case of chemical contaminations. These phenomena are not confined to the past; on the contrary, their legacy persists in effects felt over very long periods of time. The author argues that populations living in such contaminated environments are subjected to 'slow violence'. An important mechanism of this 'slow violence' is the latency between the production of pollutions and the manifestation of their deleterious effects. This latency is particularly significant in the case of endocrine disruptors. This category of substances, characterized in the early 1990s, can have effects not on the individuals exposed but on their descendants. The massive presence of such substances – namely PCBs – is thus associated with the first cluster of significant sex-ratio alteration identified in humans, namely in the Aamjiwnaang First Nation population. The second part of Murphy's analysis considers the strategies that have been developed to govern a toxic world. The paper highlights how life in such a world is now organized by a set of heterogeneous yet complementary governance strategies that have been developed gradually, in response to problems as they become too visible. These strategies seek to draw attention away from the problem, by eliminating the most spectacular effects of pollution, by adopting environmental laws focused on aesthetics and by implementing activities for 'preparedness' to major accidents – as though risk only lies in the disaster to come. Murphy also stresses how little research has been done by the authorities on contaminations and their effects. The tools developed by the government, within the framework of neo-liberal governmentality, seek to transfer the management of exposure to individuals themselves, for example by instructing residents to stay in their homes during certain periods of high pollution levels, by limiting their consumption of certain local products that have been contaminated, or by training them on the 'right' attitude to adopt in case of an accident. Murphy thus shows that these strategies, which seek both to make the problems invisible and to normalize life in contaminated areas, have allowed for the maintenance of an 'open space for business', in which polluting activities continue to develop despite the cascading and increasingly irreversible problems they generate.

The growth of these neo-liberally inspired tools is a core feature of the current management of the toxic world in-the-making. At a time when risk technologies are starting to prevail in various industries, the regulatory authorities are encountering a new type of problem head on: the lasting contamination of certain areas, to which these technologies have no solutions. The focus is therefore now on devising tools and policies for 'living with', which recognize *de facto* that the contamination is irreversible. The first illustration of these changes appears in Didier Torny's paper. Torny examines the gradual evidencing, from the late 1990s, of the generalized contamination of the French Antilles by chlordecone – an organic-chloride pesticide used in banana plantations until 1992 – and then the management of this lasting contamination for which no remedy exists. Chlordecone, a persistent organic pollutant (POP), which has a half-life of several hundred years, is an endocrine disruptor and carcinogen. Torny's analysis points out three significant phenomena in the transformation of toxic governance in recent years. First, he clearly shows the importance of the materiality of contaminants, that is, not only of their toxic effects but also of their physical-chemical characteristics, which cause these substances to remain present for a long time after their use. Second, Torny's paper underscores the existence of a 'toxic heritage' for which there is no simple solution. Regulatory authorities now have to manage not only crisis situations but also the future living conditions of populations that have no alternative but to carry on living in areas that will remain contaminated. Third, Torny's paper illuminates the shifts taking place in the production and implementation of limit values. The experts and regulatory agencies, in addition to producing these thresholds, also have to design and apply policies that compel individuals to change their practices – in this case with regard to agriculture and food – if they are to limit their own exposure. These new 'living with' policies thus endorse a certain distribution of tasks between the regulatory agencies and individuals: while the former produce limit values and recommendations, the latter have to be responsible for themselves, to know these recommendations, and to apply them.

It is the analysis of this new distribution of tasks also found in other situations of contamination, along with its implications and consequences, which is at the heart of the final contribution to the book, by Sezin Topçu. This author analyses another situation of lasting contamination: the Chernobyl accident. She examines the 'experiments' undertaken in Belarus in the mid-1990s by a group of French consultants, under the Ethos Project. Their work met with rapid success in the French and Belarusian governments, as well as certain supranational authorities such as the European Union and the World Bank. These institutions consequently extended the experiment to other territories contaminated by the Chernobyl accident and sought to draw general conclusions applicable elsewhere. The interest of Topçu's paper is twofold: first, she describes with precision the creation and implementation of these new tools for regulating toxicants, aimed above all at governing life in contaminated environments; and second, by characterizing them, she sets these tools in the context of broader changes in population governance technologies. Topçu thus shows how these practices encompass various interlinking rationalities: former thinking in terms of thresholds, aimed at controlling exposure levels; the more recent approach of neo-liberal governance in which each individual is responsible for their own health and well-being; and, finally, the participative approach, which involves the populations in policymaking. In so doing, this author brings to light all the paradoxes and ambiguities of these modalities of governance. Under cover of empowerment and participation, the management of problems is finally transferred to individuals, when in fact they bear no responsibility in the contamination they endure. This new mode of governance may seem cynical; yet, as Topçu points out, its growing success is not only due to its promotion and adoption by many institutions and governments. Some of the affected populations are actually in favour of this type of approach because they are finally offered something for 'living in' and 'living with' this toxic world that they did not produce but to which they are condemned.

Conclusion

In 1962, in the introduction to her famous book *Silent Spring*, Rachel Carson noted the extent of chemical contamination of the environment and the chain reaction of threats to the planet and to humans:

> The most alarming of all man's assault upon the environment is the contamination of air, earth, rivers, and sea with dangerous and even lethal materials. This pollution is for the most part irrecoverable; the chain of evil it initiated not only in the world that must support life but in living tissues is for the most part irreversible. In this now universal contamination of the environment, chemicals are the sinister and little-recognized partners of radiation in changing the very nature of the world – the very nature of its life. Strontium 90, released through nuclear explosions into the air, comes to earth in rain or drifts down as fallout, lodges in soil, enters into the grass or corn or wheat grown there, and in time takes up its abode in the bones of a human being there to remain until his death.[39]

When Carson was sounding the alert, numerous public policies on potentially dangerous chemical substances existed already in the US and Europe. Yet, despite the development of such regulations and several consequent changes aimed at improving them, despite the establishment of transnational systems, despite the significant development of systems of expertise, evaluation and the management of toxic substances, despite countless research studies, and despite the importance that activist movements had taken on, on all scales (local, national and global), the situation that Carson denounced in the early 1960s has constantly worsened. We are now living in a 'toxic world', in different ways, depending on the region of the world and the social condition of its population. This world is not only the outcome of toxic substances put into circulation today; it is also the result of a 'toxic legacy', the product of the accumulation in the environment, over time, of different toxic substances that may change and combine, or not. These substances, left as a legacy to future generations, even though they are banned by international conventions such as the Stockholm Convention of 2001 on persistent organic pollutants, are in some cases a source of numerous problems today. This type of toxic heritage can be manifested in its effects on the second or third generation, even for substances that can disappear rapidly, like DES or dioxins.[40]

Many solutions have been proposed, and many policies tried and tested to manage this toxic heritage and the permanent flow of dangerous substances in circulation. Analysis of the history of these policies and of the modes of management of toxic substances shows that the solution does not lie in an administrative innovation or in the production of knowledge on the effects of these substances. We already know that these effects are numerous and that they have a huge human and financial cost for our societies. But we also know that the upsurge of their production does not stem from irrationality or irresponsibility only. It is also the fruit of choices made at a particular point in time for a whole range of reasons. Yet in all these choices there is a common denominator, the centrality of economic interests. The overriding questions, to which neither REACH, nor the Globally Harmonized System of Classification and Labelling of Chemicals (GHS), nor the current revision of TASCA provide answers whatsoever, are how a number of unsaid truths about the choices made can be expressed, what the imperatives and choices of a society are at a particular point in time, and who makes the decisions. It is only by seriously examining these questions, which are above all political, that it would be possible to envisage a real, meaningful solution and not just temporary and circumstantial quick fixes to the many problems generated by toxic substances and their effects.

1 THE COLD WAR OVER THE WORKER'S BODY: CROSS-NATIONAL CLASHES OVER MAXIMUM ALLOWABLE CONCENTRATIONS IN THE POST-WORLD WAR II ERA

Christopher Sellers

Introduction

In the spring of 1959, representatives of twenty-six nations from both sides of the Iron Curtain gathered in Prague to negotiate. The Cold War was at its height: ten years before, NATO had been born, and in response, just four years prior, the Warsaw Pact. But this gathering in Prague featured neither diplomats, nor any mention of armaments or border disputes. Those who gathered in the Narodni Klub were mostly scientists, doctors and health officials. They sought, instead, to settle a striking divergence that had become ever more apparent over these same years in national approaches to industrial hazards. Specifically, they sat down to debate differences in so-called 'Maximum Allowable Concentration Levels' (MACs), the numerical limits many of these nations had established for toxins in their workplaces.[1] Behind their disagreements, ostensibly over just what levels to prescribe, lay issues of appropriate knowledge and technique; also – as will come as no surprise to readers of this book – differences of culture and power. In broad outline, the contrasts were vintage Cold War: those from the US and the USSR sharply disagreed, and the Europeans were caught in-between. In ways that STS scholars or historians have as yet hardly explored, the disagreements also foreshadowed the future course of toxics regulation not just in workplaces but across many other spheres of the industrialized world.

That this debate over workplace hazards anticipated the shift toward what some scholars have termed 'precaution' complicates our current understanding of where more preventive approaches to environmental toxins came from. Explicit statements of a 'precautionary principle' have quite recent origins, in the Europe of the 1980s and 1990s, but questions about precautionary inclinations toward industrial risks have increasingly provided analytical leverage for German, French and American historians looking at earlier periods.[2] Thus far, however, this literature has emphasized regulatory sciences devoted to the oversight of consumer or environmental exposures, such as from lead paint or auto exhaust fumes. To tell from US-based work, the mid-century experts in workplace hazards stalwartly opposed more preventive approaches emerging in these other realms.[3] Yet European examples from the earliest decades of the twentieth century, as well as those from North American, European and other workplaces later on, suggest that, under the right conditions, workplace hazards, too, could generate their own more precautionary science and politics.[4] The international dispute over MACs that stirred in 1959 Prague demonstrates how, outside the US at this moment, some groups of experts in workplace health had already become more precautionary than their American counterparts.

For this very reason, historically speaking, the mid-century debate over MACs marks an important juncture in the supra- and transnational evolution of efforts to understand and combat industrial hazards. Though the 'MAC' coinage was American, these aerial standards, also known as Threshold Limit Values, circulated widely from the 1940s onward. They quickly caught on, or were independently discovered as regulatory tools, in many other industrialized corners of the world. A variety of scholars have recently forged new models of investigation that promise a fuller understanding of this relative simultaneity. Following these, I look at historical development of supranational organizations and networks of reformers and/or experts, also by analysing national settings which gave rise to expertise which operated in similar, increasingly comparable terms.[5] I also suggest an additional way into these issues, via questions of representation forged by science studies scholars.

At issue in this debate were what science studies scholar Cori Hayden has summed up, in another context, as 'the representational capacities of knowledge, or differently stated ... the interests that ... knowledge is expected to bear, represent, or animate'.[6] The representation being enacted at Narodni was social and political. Each of the experts brought to table the historical accomplishment of those professional networks to which they belonged in their own countries, but more besides. That no companies, unions or workers, nor even state officials, sat around the table at Narodni reflected how, in each and every of these represented nations, a group of experts had achieved a certain native authority over questions of how workplace hazards afflicted workers' bodies – from bargains that I have elsewhere termed, in looking at the US, a 'pax toxicologica'.[7] Each nation's experts who spoke at Narodi thereby expressed their own nation's current configuration of workplace power relations, itself the fruit of a long social and political history. At the same time, each of those 'pax's' brought to the table had its own scientific face. Speakers clashed in their preferences for scientific argument and evidence, and pushed scientific ways of representing the interaction of workers' bodies with their environments that starkly contradicted one another.

The vulnerabilities of the more precautionary argument at this historical moment stemmed not just from contrasting socio-political bargains, but from representational dilemmas.[8] Existing clinical and epidemiological methods for studying these effects had temporal biases, against effects that unfolded over years or decades. The field of industrial hygiene thereby reflected a dynamic that shows up in Julie Doyle's study of representations of climate change, as well as other studies of difficulties of representing distant or future environmental consequences as real.[9] Even today, the long-term health effects of industrial chemicals often remain notoriously difficult to establish with any certainty.[10] In the middle of the twentieth century, as the modern scientific tools for contending with this kind of effect were just being born, the uncertainties that swirled around chronic effects of exposure were still more formidable.

It is little wonder, then, that the ways of representing workplace risks presented in Prague clashed so with one another. Counter-intuitively, the least precautionary arguments came from those whose epidemiological approach reflected the closest and most intensive contact with workers' actual bodies and working environments – the Americans. And the most precautionary arguments, by Soviet scientists who leaned on animal experimentation, involved the greatest remove from actual workers as well as workplaces. Western Europeans, meanwhile, were most skeptical of the laboratory's utility in monitoring or correcting hazards, invested as they had become in a clinical oversight that went less mediated by laboratory tests. Each way of representing and assessing effects on workers' bodies had a longer history, unfolding within as between their respective nations.

Before Prague: Bargains Clinical, Epidemiological and Experimental

Organizationally speaking, the roots of the Prague conference are directly traceable to an international group of professionals born half a century earlier, during what Anne-Emmanuelle Birn has characterized as the 'institution-building' phase of international public health.[11] Differences especially between Russian and American hygienists had first been aired and extensively debated at the 1957 meeting, in Helsinki, of the International Congress on Occupational Health (ICOH).[12] Since the early decades of the twentieth century, European precedents and leadership had dominated this and other supranational institution-building in industrial hygiene; the Americans and Russians had been relative latecomers to these discussions. Suspicion and dismissal of American approaches dated from when the US sent its first delegation to the ICOH meeting, in 1910. Alice Hamilton, the American pioneer in industrial hygiene just hired by the US Bureau of Labor, would long rankle over a remark she heard there, by a Belgian hygienist, that industrial hygiene in the United States 'n'existe pas'.[13]

That the remark overhead by Hamilton came in French was no coincidence; initially, this tongue, rather than English, had been the official language of ICOH. Though the first Congress, happening in Milan, was largely an Italian affair, this second one was dominated by French speakers, with Belgium and France having the greatest number. Even the permanent commission set up to keep the meetings going, though it strove for more even distribution of leadership, drew almost entirely from Western Europe – with the exception of Britain's Thomas Legge. The early dominance of Europeans in this organizing stemmed from the promptness with which they established full-time niches, not just locally but in their federal governments. The achieved power of labour unionism, breaking through at the national level, had secured nation-wide laws for workplace protections that included expert oversight. It was as administrators and implementers of these laws that prominent figures from Ludwig Teleky (Germany) to Thomas Legge (Britain) to Luigi Carrozi (Italy) to Marcel Robert (France) gained the authority to speak for workers' bodies in their nations' factories and capitals, and then at international forums such as the ICOH.[14] While some had university or medical school appointments, the most prominent were those with national government appointments.

Ameliorative as they saw their goals, oversight and tools, however, these were only intermittently precautionary in the modern sense, especially as these positions were consolidated and expanded within the administration of these nation-states. Around the turn of the century, in cases of phosphorus match production and also white lead paint, national legislatures did actually vote to ban substances determined to be especially hazardous to workers. France outlawed both phosphorus matches and white lead paint, and the International Labour Organization pushed similar bans in the first international standards for occupational health during the 1920s and 1930s.[15] But front and centre among the ILO's recommended standards over its first decades, and the single most widely applauded and embraced legal innovation of this period, was workers' compensation for occupational diseases.[16] The legal model was essentially reactive: workers themselves could only collect payment *after* their workplaces had inflicted a disease or accident. Not that the architects of compensation were unconcerned with prevention; a stated goal was to provide employers with financial incentives to eradicate these dangers. While an expansion of factory inspectorates also accompanied or preceded the rise of compensation laws, most of these remained far more focused on avoiding accidents than disease. In response, the more health-minded agenda of 'medical factory inspection' took shape. Starting with Belgium in 1895, its sway widened considerably over the 1920s and 1930s.[17] The next step for many European nations was to require ever more constant and ubiquitous medical oversight of workers, as a matter of law.

Even as nations such as France and Belgium continued from the 1930s through the 1940s to rely on worker's compensation for combating occupational diseases, they mandated clinics and doctors for most of their nations' workforces. Hence the gathering predominance, especially in this region, of what Rosental has described as an 'occupational medicine' model. Though slow to institute worker's compensation for occupational diseases (it only did so in 1927), Belgium required regular clinical exams in some of the most dangerous industries in the course of the 1930s. In 1945 its legislature then moved to require this medically administered 'sanitary control' of the workplace in sizeable firms across its industrial and commercial.[18] The French, while providing compensation for occupational ailments earlier, in 1919, over the succeeding decades leaned nearly as heavily on the examinations and related oversight that doctors could provide.[19] After World War II, medical oversight was mandated for most every French industrial enterprise. The explicit goal here was preventive: to remove workers from exposures that, as evidenced by required clinical exams, had already begun to make them sick. The laws also entrusted surveillance of the workplace itself to doctors, as well as *commissions d'hygiène et sécurité* which would take up causes of accidents.[20] Germany, though producing experimentation as well as shop-floor studies, by 1944 had also come to lean more and more on a clinical approach, hiring some 8,000 workplace physicians.[21] After the war, however, when the International Labor Organization took it upon itself to recommend a system of workplace health surveillance to nations, its leadership was 'largely inspired by French law and practice'.[22] The implied pax toxicologica, at once social and epistemological, relied predominantly upon harms that registered in the clinic, in signs and symptoms that caught the clinician's eye.

These European state-empowered physicians shared much with early American as well as Russian counterparts, starting with just what worker conditions or exposure were taken to be genuinely dangerous. In noted contrast with later developments, apprehension of workplace dangers hinged, whether more or less heavily, on clinical presentations, as revealed by the doctor's examination of a worker's body. The representational arbiters of hazard remained the eyes, touch and hand-held tools of the physician. In a variety of nations, some did increasingly undertake the earliest quantitative measurements of shop-floor air and blood; others conducted more and more exposure experiments with animals. But in the US as in Europe, these continued to play a subordinate role in final judgments as to whether or not a workplace was dangerous, or a worker, sick. Only when a worker's clinical exam matched a 'classic' array of signs and symptoms of poisoning or lung disease was his or her body taken to be actually damaged. Until this classic picture presented itself, workers' bodies were considered to be tolerating the environmental and bodily burdens of an occupational insult. After World War II, once the Americans' new epidemiology of occupational 'thresholds' began placing these long-unstated assumptions on the table for discussion, Europeans' pre-established orientations inclined some of them to question the quantitative and explicit thresholds set by their American colleagues.

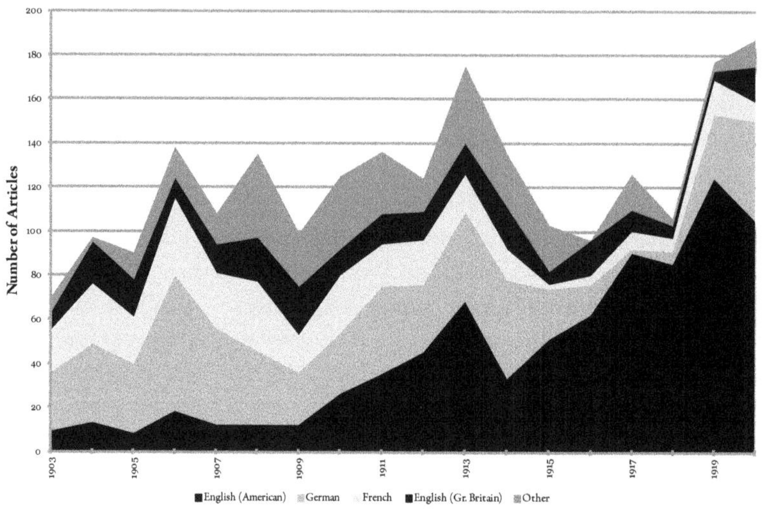

Figure 1.1: American and other entries in *Index Medicus*: showing the growth in American contributions to this US-based compilation of the literature.[23]

New Science-Building in American Industrial Hygiene

The basis for this new science, the instalment of Americans in comparably stable, professional positions to those of their Western European colleagues, began especially during the pointed labour struggles of the 1910s. For all the assumptions it shared, American institution and expert-building in industrial hygiene took different turns than in much of Western Europe. Until additional state funding was made available starting with the New Deal, the funds for America's industrial hygiene academy came from mainly private corporations, not from governments. Its advocates also faced a comparative deficit of trust in experts identified by recent scholarship of cross-national comparison.[24] American industrial hygienists responded by adopting more of the trappings of an academic science: departments, journals and career paths hinging on published research. Since nearly all the compensation and other laws for protecting workers lay in the hands of state legislatures, the federal government honed a different institutional niche. Instead of undertaking workplace inspections and enforcing laws – a federal responsibility in such nations as France, Italy, Britain and Germany – the US Public Health Service concentrated on its own programme of published, as well as strictly consultative, research.

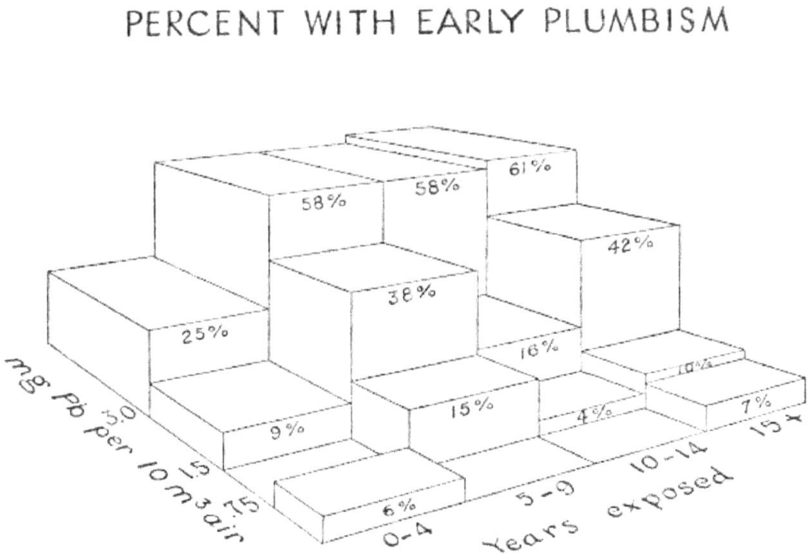

FIGURE 27.—THE PERCENTAGES OF STORAGE BATTERY WORKERS IN EACH OF 16 EXPOSURE GROUPS DIAGNOSED AS CASES OF EARLY PLUMBISM ARE REPRESENTED BY THE HEIGHTS OF BLOCKS. THUS, OF THE WORKERS EXPOSED MORE THAN 15 YEARS TO ATMOSPHERIC LEAD CONCENTRATIONS IN EXCESS OF 3 MG. PB PER 10 M.^{3} OF AIR, 61 PERCENT WERE FOUND TO HAVE EARLY PLUMBISM. DATA TAKEN FROM TABLE 34.

Figure 1.2: The PHS's epidemiological way of visualizing lead's damage to workers over time, thereby justifying a particular Maximum Allowable Concentration level, in this case, 1.5 mg of lead per 10 m^3 in the air. From the US PHS study by W. C. Dreessen, T. I. Edwards, W. H. Reinhart, R. T. Page, S. H. Webster, D. W. Armstrong and R. R. Sayers, *The Control of the Lead Hazard in the Storage Battery Industry* (Public Health Bulletin 262, 1941).

Not surprisingly, US hygienists leaped more quickly than their European colleagues toward publications, as well as scientific innovation. And if many of their private consultations and financial backers remained studiously unpublicized, over the 1920s and 1930s America's academic as well as federally-based industrial hygienists invested more time and effort in making their own knowledge claims transparent. Among the leaders in their swelling reportage of studies was the US Public Health Service (PHS) and its innovative epidemiology of the workplace.[25] Pairing aerial measurements of toxins throughout a factory with mass clinical exams of an entire workforce, it forged new ways of scrutinizing the causal influence of workplace toxins that would soon ground the American embrace of Maximum Allowable Concentration levels.

Figure 1.2 illustrates just how the results from the new PHS epidemiology represented workers' bodies as a population exposed over time. A visual correlation of atmospheric exposures with percentages coming down with clinical disease suggests how, in a predictable manner, rising exposure levels cause more workers to get sick, and vice versa. But not only does diminishing exposure coincide with a diminishing rate of lead poisoning, the rates drop off fast at 1.5 mg of lead per 10 m^3, to only a small minority of those exposed. And in such a depiction, this minority does not increase substantially even with '15+ years' of exposure. PHS investigators believed that medical monitoring of the workforce could remove this small percentage of abnormally vulnerable workers. Hence, they interpreted these results as justifying 1.5 mg of lead per 10 m^3 as the limit of 'safe exposure to lead fumes and dusts'. The goal here was preventive: maintaining this level of exposure would keep many more workers from getting sick. Nevertheless, by looking only for the clinical surfacing of classical lead poisoning, also by utterly discounting any selection bias that might have come through 'turnover' of the sickest workers, their representations of lead and other toxins' longer-term effects wrote off many consequences that others would soon elaborate.

Less purely reactive than the strategy of workers' compensation, the early use of exposure limits nevertheless restrained their own preventive potential through assumptions very much like those of the Europeans. Unlike its federally based counterparts in Europe, the PHS could only make the case for such limits scientifically. But even after other American hygienists in academe and the states had taken up and generalized this tool, these maximum concentrations remained for decades in the realm of recommendation rather than law. Alice Hamilton herself had helped publicize limits to many workplace toxins early on, through a field-defining textbook as well as her consultations with the Massachusetts Department of Labor, which became one of the first, in the late 1930s, to publish a list of recommended values. National professional bodies took over this informal standard-setting, first in 1941, the engineers of the American Standards Association, then starting in 1947, an 'American Conference of Government Industrial Hygienists'.[26]

In retrospect, these decisions about what level of toxic exposures should be allowed or considered tolerable were hardly just about scientific findings alone. They were also political and value-laden decisions, implicitly arbitrating between certainty and uncertainty, workers' health and the demands of production. That the Americans' pax toxicologica relied more on these numbers reflected the social bargains through which their own authority in the US had been bought, from employers and state health, labour officials and unions, as well as the lesser security of their expert authority. The apparent precision of the MACs, backed by the new environmental epidemiology, bolstered their clout in ways for which European colleagues saw less of a need. It also helped cement another departure with European regulatory styles: a growing reliance not just on clinicians and medical exams but on aerial monitoring, and the engineers who could undertake it.

New Science-Building in Soviet Industrial Hygiene

Conditions in that other nation which also began to lean on aerial standards early on, the Soviet Union, might seem at first to contrast starkly with those in the United States. After all, the Russian Revolution of the late 1910s had ostensibly been on behalf of that nation's workers, and the resulting nation-state promulgated an ideology that was, ostensibly, the most pro-worker of any Western nation. Starting in 1923 in Moscow, a new network of state-sponsored research institutes and training programmes were established.[27] Reflective of a very different social bargain from in the US, the state and its experts faced no private counterbalances to workers' interests, initially easing experts' access to shop-floors. As a result, when an American hygienist such as Alice Hamilton visited the Soviet Union in 1925, she found the conditions for industrial hygiene work nearly ideal: with 'no question of a sensitive or hostile employer ... [or] argus eyed insurance company'.[28] Indeed, her own later advocacy of aerial exposure standards may have reflected an awareness that, starting in 1930, the Russian Health Ministry began issuing their own set of exposure standards for factories, as a matter of nation-wide law. But Soviet Russia's enterprises, if state-run, still faced imperatives of production that could counterbalance the government's stated commitments to worker health. Not only did they still have far to go in catching up with the technology or output of Western European counterparts, the political turmoil in the late 1920s that yielded Stalinism had devastating consequences on some hygienic researchers as well as workers' own groups. The director of the occupational health institute in Leningrad, Vigorchik, had been sent to Siberia twice by 1931. As autonomous networks of labour unions were also suppressed, their demands were also subordinated to those of rapid industrialization.[29]

From the 1930s onward, Russia's industrial hygiene investigators contended with a blend of authority and impotence that was peculiar to this nation. On the one hand, they formulated a set of rules, such as the list of exposure limits that the Health Ministry could immediately translate into nation-wide law.[30] On the other hand, unlike their Western European counterparts, and more like the US PHS, they had much less hand in the Ministry's as well as others' efforts at enforcement.[31] Those at the research institutes enjoyed two productive avenues for obtaining information about goings on in Russian factories: occupational clinics, where workers from all manner of industry went, and their own laboratories. But as their human epidemiological studies remained largely confined to patients in their occupational clinics,[32] they relied, more heavily than either Western Europeans or Americans, on animal experiments as a basis for the exposure standards they promulgated.

Taking animal experiments as representative of human occupational exposures, Soviet industrial hygienists made choices less consonant with America's and Western Europe's industrial hygienists than with these nation's regulators of consumer products. At the US Food and Drug Administration and at similar regulatory authorities in France and Belgium, animal experiments played a growing role in decisions about food additives and pesticide residues.[33] Among the great advantages of animal experimentation was the measurable and verifiable information it offered on exposures that were at once minute and chronic. Even over the entire lifetime of an organism, accurate and (presumably) verifiable figures could be had on the amount of a toxin to which it had been exposed. Use of control groups and strictly maintained laboratory conditions meant that chronic effects of even single chemicals could be isolated. Autopsy and pathological and histological evaluations also allowed for the visualizing of effects that might not have surfaced as symptoms (see Figure 1.2, p. 31). Of course, the bodies involved were not human, raising many questions about interspecies variation. But compared to existing epidemiology or clinical methods, animal experimentation offered what was in important respects a more rigorous scrutiny of the chronic temporality of occupational effects.

Soviet toxicologists also parted ways with American and Western European industrial hygienists in other respects. In particular, their experimentation honed in on a part of the body in which Western industrial hygienists by the 1920s and 1930s had not shown much interest: the central nervous system. Much of this work built on the experimental tradition of Pavlov, of dog and bell-ringing fame. It was a programme well-established in Russia, but less so in the West, outside behavioral psychology.[34] For Russian hygienists, the focus on neurological damage came bound up with another departure from Western industrial hygiene. The Russians, especially in their experiments, went looking for harms that were less discernible through a standard clinical exam. This search was what justified their scrutiny of neurological effects in particular; they reasoned chemical damages would show up there first. Whereas the Americans, in particular, set allowable levels at a point where some workers in a population, even to tell from their own studies, would get clinically sick, the Russians set maximum exposure levels that included margins of safety, lower than any pathology they could actually detect. By no means did they abandon the threshold idea. What they sought was an alternative accounting of the tipping of physiological balances that defined it: preclinical rather than clinical, discernible only through laboratory and experimental means, prior to any full-blown pathology .

Russian, American, and Western European – these were three contrasting ways of speaking for workers' bodies that would clash around the table at Prague in 1959. That they could do so followed from the accomplishment, over the previous decades, of professional and official niches devoted to industrial hygiene in all these nations, also to international developments. 'Neutral' international

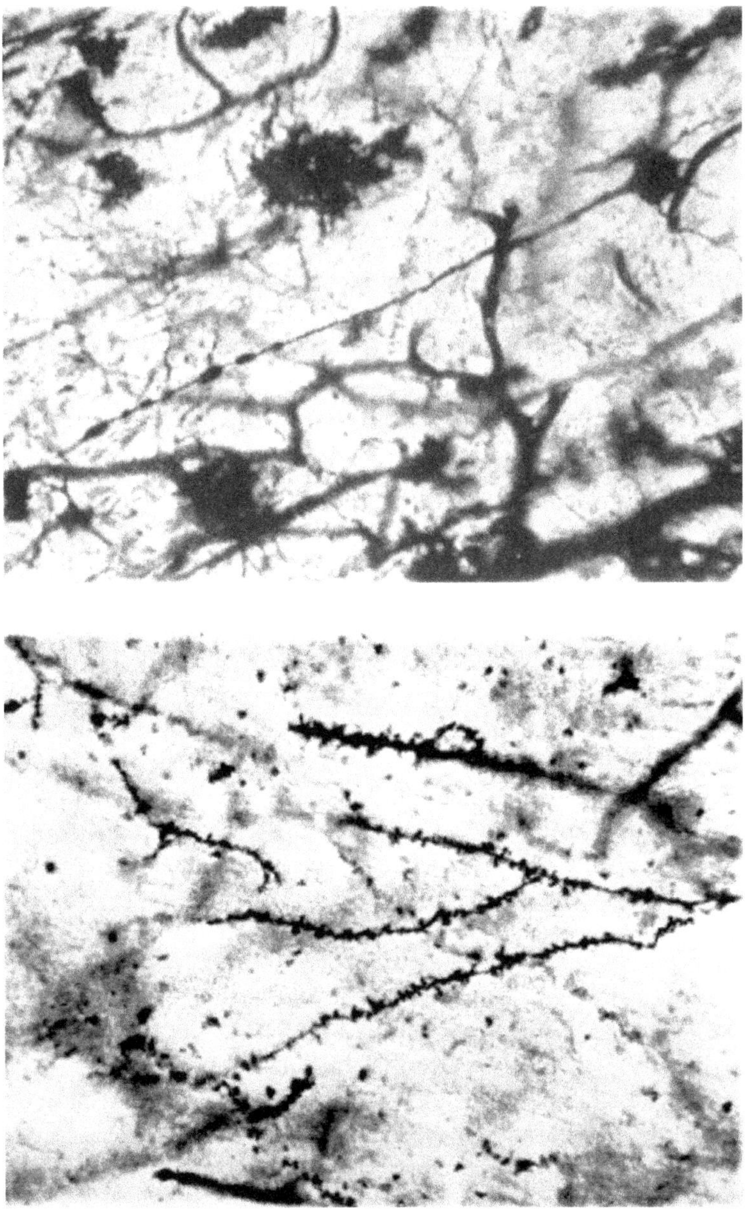

Figure 1.3: Visualizing preclinical damage over time: from a Russian exposure experiment, showing dendrites of nerve cells from an unexposed and an exposed animal.[35]

institutions joined ICOH in seeking to craft and promote cross-national knowledge and standards for workers' health. Chief among these, prior to World War II, was the International Labour Organization (ILO), which initially drew upon the same Western European leadership as ICOH, but which over the 1930s gained the US and the USSR into its membership. After World War II, it was joined by the World Health Organization (WHO). As the sheer volume of cross-national transit and interchange mounted, especially in comparison to half a century earlier, these same international organizations and meetings became riven in new ways, by the partisanship of a 'Cold War'.

The Prague Meeting: Prelude and Confrontation

When it came to Maximum Allowable Concentration levels, no individual stood closer to the centre of the swirl of post-war debate than a French toxicologist, René Truhaut. His nationality, again, was far from coincidental. The French had long been prominent in international networking for occupational health, and in the 1950s French representatives served as heads of industrial health activities in the ILO as well as International Union of Pure and Applied Chemistry (UPAC). As for the international literature, French and other Western European contributors to these subjects nearly equalled those of the Americans in the early post-war years, despite their devastated economies (see Table 1.2, p. 31). Although European scientists had been some of the first to correlate toxic exposures to physiological effects, and French toxicology had acquired a secure niche in the academy, the clinicians who dominated control strategies inside French, as well as other Western European factories, had few ways of assessing shop-floors. As one critical observer put it in 1957, the 'technical materials' for aerial measurements and other tests were 'non-existent nearly everywhere', which for him was 'proof that industrial medicine and hygiene in France are nothing but a façade'.[36] Truhaut, though more diplomatic, might well have agreed. His critical role in mobilizing an international resolution to the MAC debate stemmed, not so paradoxically, from the limited influence he and his discipline enjoyed within the workplaces of his own nation.

Trained not as a physician but as a pharmacologist, Truhaut's interest turned to workplace toxins only in the early 1950s. Then, as earlier, his contributions came in the realms of chemical analysis, animal experiments and literature reviews. His career had begun at an institute devoted to the research of cancer, the causes of which were just beginning to prompt a rethinking of threshold assumptions in many corners of the scientific world. Even as he began to investigate, unlike his American counterparts, he continued to study and write about the dangers of food additives and pesticides, realms in which, in the US as well as Europe, more precautionary methods reigned.[37] He helped engineer the Prague confer-

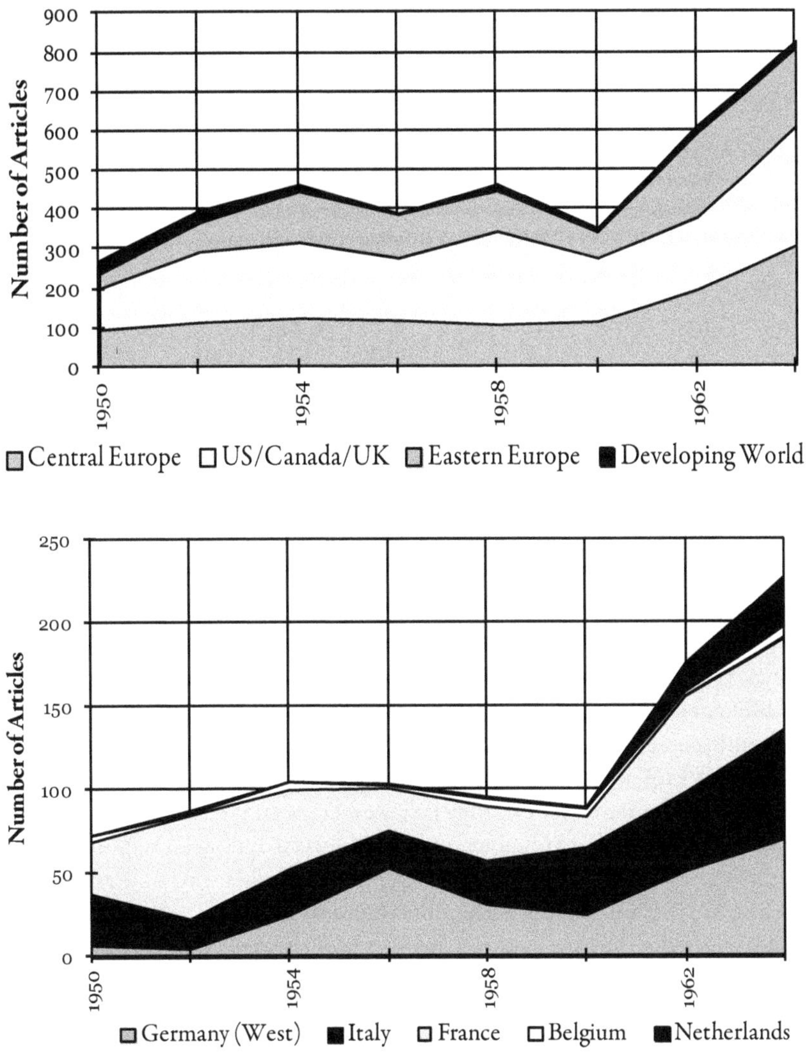

Figure 1.4: Post-war industrial hygiene publications by region and by Western European nation. From *Index Medicus*.

ence after attending the 1957 meeting of the ICOH in Helsinki, which included a session featuring several speakers on 'Industrial Hygiene Norms'. American, Argentine and Russian presenters made plain the stark differences between the 'norms' of different nations, in particular, between the maximum concentrations being promulgated by the Soviets versus the Americans (see Table 1.1).[38]

Table 1.1: Differences in MACs reported by a German researcher in Helsinki, 1957: *Schwefelkohlenstoff* is carbon disulfide; *Blei* is lead; *Quecksilber* is mercury. From ICOH, *Proceedings of the XII International Congress on Occupational Health: Helsinki, 1.–6. 7. 1957* (Helsinki: Valtioneuvoston kirjapaino, 1958).

Stoffe	Maximalzumassige Konzentrationen in Mg/L	
	UdSSR	USA
Anilin	0.005	0.019
Schwefelkohlenstoff	0.01	0.06
Kohlenstoffoxyd	0.03	0.11
Dichloraethan	0.05	0.4
Chlorkohlenstoff	0.05	0.13
Trichloraethylen	0.05	1.0
Blei	0.00001	0.00015
Quecksilber	0.00001	0.0001

Struck by 'the sometimes considerable differences between the values adopted by different countries', Truhaut helped spearhead the organizing of a symposium to examine and debate these differences, with financial and other support from IUPAC, WHO and the ILO.[39] In terms of numbers, the national representatives who showed up were carefully balanced between the two sides of the Iron Curtain, while also including experts from neutral European nations. Representatives of NATO nations, which had joined together into military alliance starting in 1949, formed the biggest bloc, but only by a small margin. The Czech delegation, from the host nation, a member of the Eastern Bloc, had the biggest group. But the US and the UK joined Italy in sending the next most representatives; followed by large representations from Russia and the Netherlands, and smaller ones from other European nations. Neutral nations – Switzerland, Sweden, Spain and Japan – made up the balance.

If the representatives gathering in Prague reflected the parallel rise of industrial hygiene expertise across many nations after World War II, many at the Prague conference spoke of the MACs as having a largely American vintage and imprint. We may understand this neglect of the Soviet experience as partly due to the relatively more publicized and transparent American scientific justifications of their values, but it can also be attributed to Cold War biases. In a way, however, the Soviets' emphasis on laboratory experimentation with animals anticipated where the Americans themselves were already headed. Experimentation with animals, and for less deadly exposures, with humans, did occupy an increasingly pivotal role in the American evaluations by this time. Among the fifty-two gases and vapours for which these were estimated in the 1958 MACs, about 56 per cent of the total included animal data, that is, from experimental laboratories. The Americans preferred British or American experimental reports, whose emphasis on pathology and biochemistry, as one British observer admitted, 'would seem to have little to offer' when it came to nervous system effects. But human epidemiology, especially if carefully correlated with measurements of exposure, was determinant. The Americans only based eight of these fifty-two standards on animal data alone.[40]

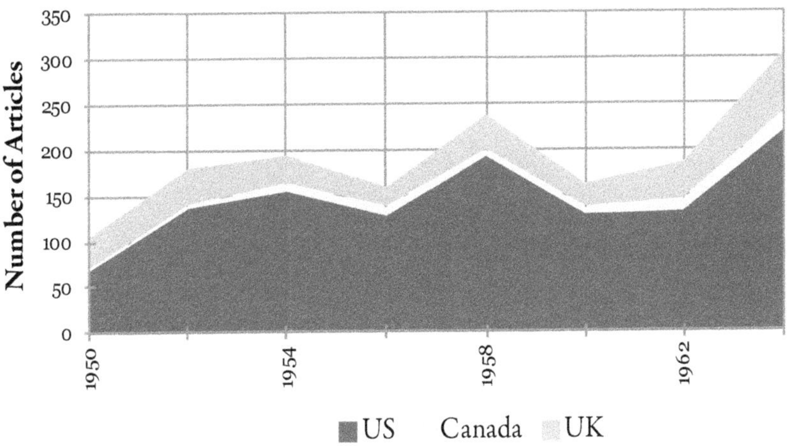

Figure 1.5: English-language publications in industrial hygiene. By the post-war era, American publications were by far the most numerous in the English-speaking world.

Over the 1950s and into the 1960s, the Russian experts, with their quite different scientific approach and practice of standard setting, emerged as the dominant national group in the Eastern bloc (see Figure 1.6).[41]

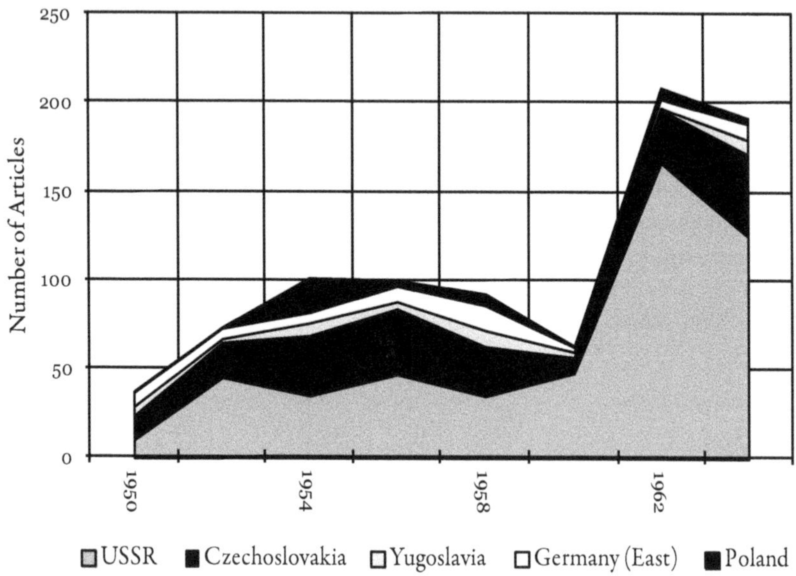

Figure 1.6: Soviet-bloc publications in industrial hygiene. This diagram shows the initial post-war prominence especially of Czech publications, but then the growing dominance of Soviet-published articles.

At the Prague meeting, the Soviets' laboratory experiments with animals figured far more heavily in their recommended safety limits than in those of the Americans. For the Soviets, experimentation was the starting point, for which 'medical examination of the workers ... in an atmosphere filled with harmful substances, and other clinical and statistical data, can serve as material only for the correction'. They repeatedly emphasized ambitious goals for their experimental designs: capturing long durations, even lifetimes, of exposure. Seeking 'highly sensitive methods', which would 'register even minor changes in the organism of the animal', was precisely what drew them to the 'nervous system' that Anglo-American toxicology by-passed, where 'the first signs of the influence of various substances are shown'. During the exposures, they checked rats' ability, for instance, to 'press against a glass door to get food', and hooked the animals up to electro-encephalograms. Their experimenters also told of other methods that were unusual among Western toxicologists. To enhance the effects to toxic exposure, they would subject animals to food deprivation at the same time, or force them to become tired through swimming beforehand.[42] Seeking techniques to highlight detrimental effects that would otherwise not be observed, the Russians and their East European allies were, in effect, biasing their science toward lower safety limits than their American counterparts. For these reasons, and because of their commitments to safety margins, their limits for many chemicals were twice to twenty times lower than their American or Western European counterparts.

If across the West, industrial hygiene experts had carved out ways of speaking for workers' bodies, in place of the voices of either employees or employers, the differences in American versus Soviet threshold numbers pointed to just how ideologically tinged this expertise remained. Clues as to just whose vision of the worker's body the American standards expressed can be taken from how Herbert Stokinger, chief toxicologist at the US PHS and one of its initiators, introduced the effort to his American industrial hygiene colleagues. As a 'keynote of enlightened industrial medicine', he declared as an article of faith that: 'No substance is so toxic that it cannot be used if sufficient knowledge of its action has been made available; similarly, no substance is so nontoxic that it should be used without regard to caution'.[43] With these last words, Stokinger was himself quoting from the medical director of Eastman Kodak, a major chemical manufacturer. Profoundly committed to making any and all toxic chemicals work for employers, the American hygienists simply assumed that this commitment could always coexist with a cautious shielding of employees' bodies. Russian hygienists, on the other hand, were under considerably more pressure to produce official standards that matched their government's official pro-worker ideology. In formulating nation-wide standards, and in their capacity to ban some chemicals, they sought a science of industrial hygiene, whose representations of workers' bodies were, at least ostensibly, more precautionary and protective. But as a science based in the laboratory and animal experimentation, its remoteness from actual shop-floors and worker exposures also roused the skepticism of Europeans, who were accustomed to less quantified and more clinically based methods of oversight.

European versus American Responses to Soviet Science

By the time of the 1959 conference, the most pointed statements of opposition to the idea of the MAC itself came not from the Russians, but from Western Europeans. At Prague as well as elsewhere, the head of Belgium's medical factory inspectorate, A. Uytdenhoef, waxed at length about the 'more or less serious errors' that attended the MACs' application. The single number given as a MAC for benzene, for example, offered little handle on how, in the petro-chemical industry, benzene was most often mixed with other chemicals which had other toxicological effects, in widely varying proportions. And what about the MACs for mineral dusts, which included no mention of particle size, a vital determinant of what gets into the lungs?[44] A Dutch government physician was still more scathing: 'the concept of maximum permissible concentration' not only had 'little validity' it was a 'dangerous concept, and based on a misunderstanding'. Significantly, these attacks drew little on animal experimentation, and far more on these experts' invocations of workplace realities, including worker medical exams. The Dutch representative did propose a version of the MAC that he thought more viable. But instead of a single number for each toxin, his alternative would have included the entire range of limits proposed for each toxin, from around the world (see Figure 1.3).[45]

As for the Americans, in the late 1950s the conditions which they witnessed in Prague and elsewhere made it easy for them to slough off the Russian and East European results. Researchers from these nations, for one thing, presented fewer details about the conditions or results of their experiments, to such an extent that the American Harriet Hardy found them 'without any supporting data'. Hardy's suspicions were also roused by the course of discussion itself, 'with wonderful equipment for interpretation into the five languages of the meeting', but far less freedom of expression than Hamilton had hopefully recorded in the mid-1920s. 'Every time someone from the US Canada or the UK spoke, his offering was greeted with silence, whereas when anyone from the Soviet Union spoke, there was enthusiastic clapping.' While she found the 'well-informed' Russians to have 'done some original work', she felt 'their whole plan was to turn the meeting to their political advantage'.[46]

What a team of American hygienists later learned, when invited in 1964 on a hygienic tour of Soviet research and industrial facilities, offered a more mixed picture of Soviet accomplishments, beyond the realm of the laboratory. Soviet hygienists were indeed testing factory air, but using methods some of the Americans called into question. They did carefully reduce exposures to many toxins, and even banned some, but were surprisingly lax about the risks of accidents. The Americans also carefully noted that they only saw a select few Soviet factories, likely the best-run.[47] After the fall of the Berlin Wall, the actual and historical conditions of factories in the USSR turned out in many cases to be horrendous. Workers at the first USSR nuclear processing plant at Mayak have since served as a unique epidemiological cohort, since it is clear that, especially from 1948 to 1958, they were 'exposed to levels of inhaled plutonium much higher than those in other countries'. Elsewhere, the evidence has also mounted that, even after the Stalinist era, pressures of production often trumped worker as well as environmental protection.[48]

And yet, those scientific assertions made by Soviet hygienists in Prague in 1959 had their own transnational history. Entailing new, more searching and precautionary representations of how workers' bodies responded to toxins over time, they bolstered the emergence of scientific communities and claims elsewhere that broke through assumptions undergirding ways of representing workplace harms that prevailed in other nations. Though his American colleagues initially found it easy to dismiss their claims, Truhaut, for one, did not. He helped engineer further meetings and symposia over the next decade that sought to resolve these international differences, in part through discussion

Part 1. Scale of toxicity

Class	1	2	3	4	5	6	7	8	9	10
Concentration in p.p.m.	< 0·1	0·1–1·0	1·0–10	10–50	50–100	100–250	250–500	500–1000	1000–3000	> 3000
Name of product and chemical composition	Figures in brackets in p.p.m. v/v									
X				A_1 (……) A_2 (……)		B_1 (……)	B_2 (……)	C (……)	D (……) E (……)	F_1 (……) F_2 (……) F_3 (……)
Y			A_1 (……) A_2 (……)		B_1 (……) B_2 (……)	C (……) D (……) E (……)	F_1 (……)	F_2 (……) F_3 (……)		

Part 2. Other information

	Recommended safety factor					
1	2	3	4	5	6	7

A_1 and A_2 the highest and lowest M.A.C. values suggested in the literature by some authoritative body.
B_1 and B_2 threshold limit (man) (highest and lowest value)
C Early signs and symptoms (man); quantitative limits of signs: *e.g.* haematoporphyrinuria, trichloraceticacid, *etc.*
D Disease (signs and symptoms to be mentioned)
E Serious disease with loss of working capacity (signs and symptoms to be mentioned).
F_1 Lethal poisoning (by chronic exposure)
F_2 Peak concentration causing serious disease (by acute exposure)
F_3 Peak concentration causing death (by acute exposure)

1 LD diagram (to mention types of animals used)
2 Recommended safety factor for " normal " people
3 Recommended safety factor for sick people (type of illness to be mentioned)
4 Recommended safety factor in case of oversensitiveness (allergy, or for combination with other toxic substances (*e.g.* alcohol)
5 Some physical data
6 Recommended safety limit
7 Special remarks and basic literature concerning figures given in the list

Figure 1.7: Proposal for each MAC as range of values, by G. C. E. Burger, a Dutch expert, in collaboration with Dutch officials.

of 'Methodology for Functional Exploration of the Central Nervous System'. When the ILO published its second edition of the *Encyclopedia of Occupational Health*, in 1971, it included both values for every toxin.[49] With so many Europeans and international initiatives taking the Soviet science and standards seriously, Americans also took greater notice.

Aftermath

By 1972, Theodore Hatch, the American hygienist and engineer, could summarize this Cold War over the worker's body in terms that cast the Soviet case in a more sympathetic light. The difference was not so much one of biases or evidence, but of 'basic concepts'. US hygienists had started by taking 'no serious threat to health ... to exist as long as the level of exposure does not induce ... a demonstrable disturbance of a kind and degree that is accepted as an indication of potential sickness'.[50] The Soviets, on the other hand, had begun 'from the opposite pole'. They took a 'potential for ill health to exist as soon as the organism undergoes a change of any kind from its normal state'. While Hatch exaggerated how far the Soviets had trod along this path, and not just in shop-floor implementation, he was nevertheless giving the Soviets their historic due. In the mid-twentieth century, when it came to workers' bodies, the Russians, rather than the Americans, had defined the more precautionary position. It was one with which American researchers and regulators – and now also union industrial health advocates – were just catching up. They were able to speak and act anew on behalf of workers in part because of the mounting conclusions about workers' bodies, as scientific objects, which Soviet research, as well as rhetoric, had anticipated. 'Increasingly sensitive tests for preclinical, physiological, biochemical, and functional disturbances ... [had begun] to demonstrate toxic effects resulting from lower and lower levels of exposure' (see Table 1.6).[51]

Historically speaking, precautionism has been something of a moving target, defined in relation to contemporary and usually pre-existing opposites. In the United States, the precautionist momentum that Hatch noted in 1972 would continue to build, courtesy of many factors internal to the US: a growth in government research funding, as well as new federal environmental and consumerist regimes, to regulate not just occupational but an entire range of other industrial hazards. By 1980, for instance, the American scientific understanding of lead in particular, as well as carcinogenic substances, was lurching in directions that Soviet science of the 1950s had foreshadowed. The new understanding of lead that would propel much of this shift was, in the American context, less about workers' bodies than those of children, and the effects of childhood lead poisoning from house-paint. But as the US Centers for Disease Control's recommendations for the allowable amount of lead in blood slipped down from the 1970s onward, the emphasis fell on just those neuro-behavioural effects that the East Europeans had studied. Similarly, substances established as cancer-causing, like benzene, began to get much stricter treatment.

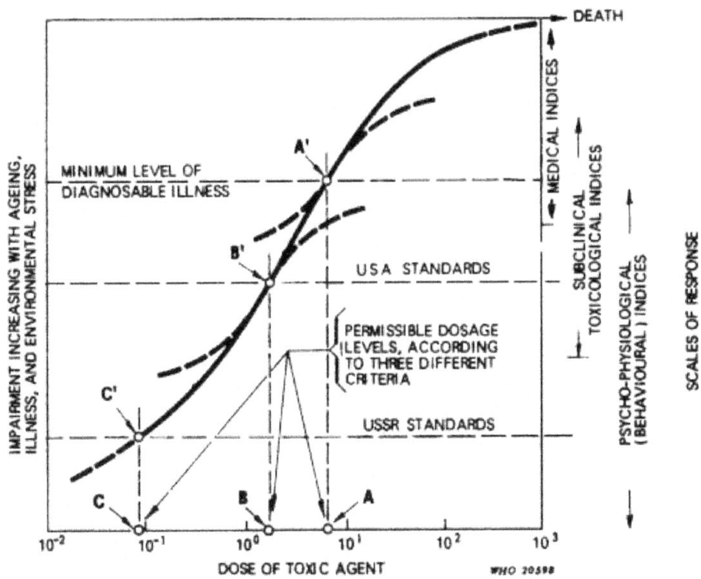

Figure 1.8: Idealized depiction of differences between Russian and US 'basic concepts' of MACs, from T. F. Hatch, 'The Role of Permissible Limits for Hazardous Airborne Substances in the Working Environment in the Prevention of Occupational Disease', *Bulletin of the World Health Organization,* 47:7 (1972).

By no means do I mean to deny the innovations or initiatives of those American researchers, policymakers and activists who propelled this precautionary shift. Certainly many of the scientists especially were aware of the considerable European support for the positions on behalf of workers' and others' bodies that they staked out in the American context. But behind these precautionary turns both in the US and in Western Europe lay the Russian and East European scientific challenges of the 1950s. They too look to have contributed to the making of a new, more precautionary approach to such hazards across the industrial West. If I am right, then these shifts were not as exclusively consumer- or extra-workplace oriented as many of the current interpretations conclude. The official ideological allegiances of Soviet and allied researchers, more avowedly on behalf of *workers'* bodies and presumed interests, had long since translated into a more precautionary science. It had done so in the nations of the Warsaw Pact , via the ways that industrial hygienists there worked in their laboratories, if not in their factories. So when precautionist factions did begin to win out in the US, it was not just a national but a transnational story, and one into which a re-imagining of workers' bodies had also fed.

2 ADAPTING TO 'REALITY': THE EMERGENCE OF AN INTERNATIONAL EXPERTISE ON FOOD ADDITIVES AND CONTAMINANTS IN THE 1950s AND EARLY 1960s

Nathalie Jas

Introduction

On 1–5 November 2010, the World Health Organization (WHO) and the Food and Agriculture Organization (FAO) held a Joint Expert Meeting in Ottawa, Canada, with the support of the European Food Safety Authority (EFSA), Health Canada, the US National Institute of Environmental Health Sciences and the US Food and Drug Administration (FDA). The purpose was to 'review toxicological and health aspects of Bisphenol A (BPA)'.[1] This joint meeting was one of the many events marking a controversy that started in the US and spread to many Western countries and international organizations from the early 2000s. Some scientists and environmental health advocacy organizations are of the opinion that exposure to very low doses of xenoestrogen, which belongs to the new category of endocrine disruptors, can cause serious diseases. The fact that BPA is present in food has prompted anti-bisphenol activists to call for a ban. Regulatory agencies and industry argue that BPA is harmless 'when used as intended' and that populations' exposure through contaminated food is 'well below levels considered harmful'.[2] The conclusions of the experts at the joint meeting were consistent with those of the firms producing and using BPA, which disregard any arguments put forward by scientists and environmental health movements. The latter group do not consider the tests implemented in the evaluations of the expert committees as 'science'. They claim that science is bypassed altogether. These tests are allegedly unsuited to the new issue of endocrine disruptors because they cannot be used to evaluate the danger of a substance in very low doses, a substance which may be active at certain stages of development and can have multiple end points and trans-generational effects. On closer inspection, we see that the controversy over the sanitary effects of this food contaminant, BPA, essentially concerns the tools used for the official expert assessment, and what they make possible to evaluate or not. Because these tools are believed to exclude de facto a whole range of problems and issues, some critics argue that they are designed to promote industrial interests over public health interests.[3]

As the BPA controversy reminds us, the tools that are mobilized determine the framing of expert assessments. They determine *ex ante* the questions to which an assessment can or cannot provide answers, and hence the problems that can or cannot be investigated. Understanding the results of expert assessments therefore requires an analysis of the construction and the choice of the tools mobilized, with a particular focus on what those tools allow for or not. Yet, even though the production of expertise has for a long time been a research subject of STS (Science and Technology Studies), the construction and choice of the tools of expert assessments, and the consequences in terms of the characteristics of the conclusions reached, has not received the same kind of attention. Two lines of analysis have largely structured STS work on expertise. First, studies have been designed to understand the social, cultural or institutional engagements and practices that determine scientific expertise and decisions within policy frameworks.[4] Second, they have sought to establish conditions leading to more scientifically valid decisions by different regulatory systems.[5] With some exceptions[6] STS research has rarely investigated the construction of the regulatory tools and the tools for assessment as well as the framing of expert assessments and of their results by the tools used.

While the BPA controversy has played out at national level, the convening of the joint meeting in Ottawa, with the support of a European agency, shows that both the UN and the EU consider the public health problems caused by food additives and contaminants, as well as their political and/or economic consequences, to be significant issues at international level. These organizations have been involved in food safety issues for over half a century, since their inception in the early 1950s. Their involvement resulted, among other things, in the early development of international expertise on 'substances intentionally and unintentionally added to food': food additives and food contaminants.[7] Although national and regional regulatory systems were instrumental in the production of regulations aimed at ensuring food safety, the regulatory tools produced by international expert committees also played an important role. Apart from the fact that international regulations have exercised forms of constraint on the regulations produced at national levels via the Codex Alimentarius system, among others, some of the tools used in expert assessments and elaborated within them constituted references that circulated in the national regulatory agencies, which were thus able to adopt them.

I would like to revert here to the emergence of this international expertise on food additives and contaminants. This story is significant for at least two reasons. First, it shows that substances – in this case carcinogens – that do not obey the foundations of classical toxicology according to which 'the dose makes the poison' have been a challenge for regulatory bodies since the 1950s, long before the issue of endocrine disruptors surfaced. Second, and following on from the previous point, the way in which the emerging international expertise on additives and food contaminants addressed the issue of the effects of low doses of carcinogens in the 1950s and the early 1960s put in place dynamics that are still currently entrenched, and found in the current controversy surrounding BPA. Thus, from the time of its creation and within that creation, this expertise integrated what the first international experts called 'reality', that is, economic development. They deliberately developed tools for expert assessments which, rather than being based on what they called 'theory', that is, state-of-the-art science, had 'practical' aims, meaning they could comply with the constraints imposed by industries. The capacity of these first international experts to adapt to 'reality' probably enabled them to ensure the longevity of the expert committees on which they served, which often had a precarious status. However, this had a price, for it required them to successively give up what they considered to be the principles and tools of expertise aimed at protecting public health.

The New Dangers of Carcinogenic Substances

In the late 1940s the old issue[8] of the regulation of substances intentionally or unintentionally added to foods was once again on the agenda.[9] Several countries took steps to review their legislation and for the first time the subject permanently became an issue at international level. Yet, despite the role of the Joint FAO/WHO Expert Committee on Food Additives (JECFA), created in 1955, in the international regulation of food additives and contaminants, and of the Joint FAO/WHO Meeting on Pesticide Residues (JMPR), created in 1963, the earliest international expertise in this field was not initially a result of joint action by the FAO and the WHO. At first it was spawned by another dynamic that was largely initiated by German toxicologist-doctor Hermann Druckrey and French toxicologist René Truhaut. In this respect, one of the main players was the Deutsche Forschung Gesellschaft (DFG, German Research Council), the only national organization to have been mentioned as a reference by the WHO/FAO joint conference on Food Additives, which in 1955 proposed the creation of what became the JECFA.[10]

It was via the burning issue of carcinogenic substances that Druckrey and Truhaut projected the regulation of food additives onto the international scene. Although Herman Druckrey was a gifted and renowned researcher, his strong involvement in the Nazi regime had limited his post-World War II career in Germany. In the late 1940s and early 1950s he acquired an international reputation after publishing the first mathematical model on the low dose-effect relationship of carcinogenic substances, based on his research on 'yellow butter', an artificial food colouring known to be carcinogenic.[11] This model served to spread the idea that, for carcinogenic substances, 'a threshold dose, which can be regarded as harmless, does ... not exist. Contrariwise, even the smallest doses are dangerous, if they act over a long time.'[12] Druckrey thus profoundly challenged the belief underpinning toxicology and the regulation of toxic substances that 'the dose makes the poison' and that it was therefore possible to establish thresholds below which no significant harmful effect could be found. In a context where synthetic chemical substances were proliferating and their use was spreading rapidly, the prospect of 'a limited exposure [being] sufficient'[13] to trigger a carcinogenic effect worried the specialists. It was this concern that Druckrey and Truhaut fuelled and mobilized to try to develop international expertise on food additives and contaminants. To this end, they used first the DFG and the International Union against Cancer (UICC).

According to Druckrey, from 1949:

> In the conviction that an effective campaign against cancer in the long run can be achieved only by prevention ... the DFG entrusted a number of scientific commissions of laboratories and industries with the task of creating the basis for scientific exploration for the protection of people against harmful and unhealthy, especially carcinogenic, substances in the environment and food.[14]

Considering that

> [t]he extent of the necessary tasks of exploration is beyond the means of one country ... the DFG during the International Cancer Congress in Paris in 1950, let it be known that it was their desire to achieve international collaboration in this field. Furthermore, they also asked, in the name of experts from other European countries, the UICC to found an 'International Cancer Prevention Committee', to consider practical efficient measures for the prevention of cancer at the 6th International Congress in São Paulo [scheduled for July 1954]. [15]

'With the intention of starting the necessary preparatory work', the German organization financed a conference for 'experts' from ten European countries. This conference, of which the kingpins were Druckrey and Truhaut, took place at Bad Godesberg on 1 May 1954 under the chairmanship of Nobel Prize laureate Adolph Butenandt.[16] Even though the 'Godesberg Conference' covered only food additives, with a particular interest in artificial colouring, and not all carcinogenic substances, it had considerable influence. It resulted in a set of proposals on international regulation of food additives, known as the 'Godesberg proposals', which subsequently became unavoidable references. They articulated expertise developed by Druckrey and the DFG, to proposals made by Truhuat two months earlier, in March 1954, at the meeting of the Committee of Experts for Sanitary Control of Foodstuffs of the Brussels Pact.

In July 1954 Druckrey and Truhaut attended the São Paulo congress of the UICC, at which the International Cancer Prevention Committee that the DFG had called for was effectively created. Even though 'cancer prevention' concerned fields other than food additives, this issue prevailed in the discussions chaired by Druckrey and Truhaut. A panel on 'regulations governing dyes and other additives in foods' was even organized.[17] Druckrey, who was both the chairman and the moderator, and Truhaut, who delivered a long paper, used it to present and discuss the Godesberg proposals and the position that the UICC had to adopt in this respect. The advantage of the UICC was that this organization provided Druckrey and Truhaut with possibilities for interaction that were far greater than those afforded by the Bad Godesberg conference. The 'International Cancer Prevention Committee', consisting of five Americans, six Europeans and one Brazilian, was chaired by a well-known cancer prevention specialist from New York University, William Smith. The WHO and the FAO were also represented by their 'scientific experts' in charge of this subject, respectively Dr Chapman and Dr Van Veen. The most noteworthy member of the panel, apart from Druckrey, who was presented as a member of the DFG Food Dye Commission, and Truhaut, who was introduced as a member of the 'Committee of Experts for Sanitary Control of Foodstuffs established by the Pact of Brussels', was the chief of the Environmental Cancer section of the US National Cancer Institute, the American Wilhelm Hueper.[18] Truhaut and Druckrey pursued their action within the UICC by organizing a 'symposium' in its name, on Potential Hazards from Chemical Additives and Contaminants to Foodstuffs. This symposium, held in 1956 in Rome, was attended by fifty scientists from twenty-one countries, as well as representatives of the WHO and the FAO. At least two-thirds of the experts participating in the six working groups were familiar with international meetings in which the topic of food additives and contaminants was starting to be discussed.

From 'Ban' to 'Positive Lists': Constructing 'General Principles' of an International Regulation of Food Additives

The minutes of the debates that took place at the IUCC congress in São Paulo and at the Rome symposium were highly detailed – unlike those of the Godesberg conference which were brief. The reasoning underlying the drafting of the Godesberg proposals is therefore easier to understand. Ideally,

> the so-called method of exclusion or total prohibition has the advantage of leaving foods in their natural unrefined form ... It is moreover surer, because it is very difficult to prove the harmlessness of a substance, and then to carry out a very long and tricky

> experiment on animals, of which the results cannot be applied de facto to humans, due to the importance of the species factor in receptivity to poisons in general and to carcinogenic substances in particular.[19]

Truhaut would have liked to adopt this strategy but feared 'that the systematic banning of all foreign substances, especially those with useful effects, without evidence of their toxicity, would not be complied with by firms, which would then make their own choices of substances without toxicity'.[20] From a 'practical' point of view, this position seemed therefore to go 'against disease prevention that was so desirable'. The necessity to take into account 'reality', that is, the weight of industrial interests and economic development, resulted in the development of another strategy to promote: the banning of all substances that have not been authorized. This amounted to producing 'positive lists' of authorized substances recognized as being harmless and especially non-carcinogenic. Correlatively, no substance recognized as having irreversible effects – and this was a key point – and especially carcinogenic effects, was to be authorized in food. Other principles were added to these two main general principles: authorization was to be granted to substances only in so far as 'there should be an actual need for their use'; that such use would not cause the consumer to 'be misled as to the true value of the food', and that 'the amount of substances used should be as small as possible'.[21]

The decision to opt for positive lists rather than banning was the first sign that these international experts were backing down. If the so-called positive list method was to protect public health effectively, it would be necessary to develop scientific expertise capable of showing the danger, or not, of each of the numerous additives used. Other problems also had to be studied. The question of the purity of the authorized substances, that is, the absence of impurities such as arsenic and heavy metals, was considered essential. The development and standardization of reliable toxicological tests were another important requirement. Truhaut, Druckrey and the other European specialists present at Bad Godesberg were of the opinion that solving all the problems that positive lists entailed required complex and costly work. In view of the constantly growing number of substances used, such work could only be undertaken in the framework of international collaboration between the specialists concerned, allowing for the sharing of information and a division of the work. This collaboration started to be organized between European experts outside of the main international organizations.

Developing International Expertise

While the UICC was a locus of the promotion of what the international regulation of food additives ought to be – as defined by the experts that Druckrey and Truhaut had convened at Bad Godesberg – the actual tools of international expertise started to be elaborated within the standing European Committee for

Research on Chronic Toxicity Hazards. This committee, chaired by Truhaut, and which subsequently came to be known as Eurotox, was a direct result of the Bad Godesberg conference. As a follow-up to that conference, a second meeting, chaired by Truhaut, had been convened in Wageningen in July 1956. The scope of the latter meeting was greater than the former, in so far as WHO and FAO scientific experts in charge of these issues also attended. On this occasion 'it was decided to extend the field of activity to the toxicity hazards resulting generally from the addition of extraneous substances (the so-called food additives) to foodstuffs, and to hold similar meetings annually in the future'.[22] A third meeting, chaired by J. L. Dols from the Netherlands, who was one of the key players in international expertise on food additives until his death in 1980, took place in April 1957 in Ascona, Switzerland. It was at this meeting that the standing European Committee for Research on Chronic Toxicity Hazards was officially created. This committee was presented as 'an independent association of scientists for research into, and protection against, chronically toxic exogenous hazards, in the general interest' whose 'purpose' was

> to provide a scientific basis for the protection of public health and to put its findings at the disposal of the WHO and the FAO as well as all international, European and national legislative authorities and other Services which may need them.[23]

The activities of this committee, which revolved around food additives and contaminants, were a focus of keen attention. These meetings were attended by scientific representatives of the WHO, the FAO and European experts from universities, research institutes, legislative authorities and industries.

The published minutes of these meetings[24] bear witness to work geared towards producing tools for expert assessments, ranging from 'general principles' to recommendations regarding 'procedures for toxicological experimentation', as well as a series of first 'positive lists'. As had been the case at the Bad Godesberg conference in 1954, the 'European experts' that were present adopted a first list of fourteen colouring agents that they considered to be harmless and therefore fit for authorization (the Proposed European List of Food Colours). This list was based on a list produced by the DFG. Although Druckrey, Truhaut and the other experts were confident about their choices, many substances present on this list had not undergone all the toxicological tests. Yet the resolutions that they had just adopted affirmed the absolute necessity of the tests in order to guarantee the innocuousness of a substance. The list was nevertheless supposed to be proof that international expertise could be effective and produce tools of use to national and international regulatory agencies. The Eurotox committee then worked on compiling other lists, in particular four European Positive Lists of Preservatives (antimicrobials and fungicides). The first consisted of substances that could be used 'without special limitation of quantity'; the second, substances that 'have been used for food preservation frequently and for a long time whose properties, however, cannot be considered sufficiently investigated'; the third, substances whose use was 'in the present state of knowledge not recommended'; and finally, the fourth, a list of substances 'whose properties were not sufficiently known'.

Research programmes involving experts from various countries were launched, such as the toxicological research programme on issues like the 'dose-time response relationship in different carcinogenic substances', or 'the characterization of the action of co-carcinogens of promoters in order to assess the danger of their presence in the human environment'.[25] A large-scale programme focused on the development of methods and 'the possibilities for the development of standard conditions for animal experimentation concerning, for example, the choice of adequate animal species and strain, diet, dosages and route of application', 'the possibilities for limiting the toxicological claims to a practical extent and for simplifying the methods'. Another programme studied the 'various substances present in food'. Although short, the minutes tell us a great deal about these European experts' concerns and two issues in particular. First, they wanted to understand the long-term effects on health of repeated exposure to low doses of chemical substances. From this point of view, they focused closely on carcinogenic substances but sometimes also investigated mutagenic effects. Several problems seem to have interested them, especially interactions between substances, substances which are not carcinogenic but can also promote cancer, or carcinogenic effects on the descendants of exposed individuals. Second, they wished to define what could be a good toxicological test, that could be used to highlight the toxicity – in this case carcinogenic effects – of a substance, and that could serve for an expert assessment.

In spite of the dynamic approach of the members of this committee, especially its most active ones (Druckrey, Truhaut and the 'Dutch Group'),[26] the difficulties encountered by the committee in its work were evident. First, the lack of financial and material resources was obvious. The minutes regularly underscored both the long and costly nature of the research that was required, and the lack of facilities to undertake such studies. Appeals were repeatedly put out, urging states and international organizations to invest in research and to fund research programmes on the long-term effects of chemical substances, especially food additives and contaminants. Second, these experts lacked scientific data. While the European cooperation that these scientists were trying to set up was intended to offset the lack of data by facilitating information trading and data production and organizing the collaboration of several countries to produce data, in order to avoid duplication and useless research, access to FDA data seems to have been a major issue. It therefore appears that these experts saw the committee as a means to attempt to gain credibility in the eyes of FDA scientists. The third difficulty was the absence of institutional anchorage. This meant that the committee produced international (European) expertise that had no immediate application, and that it would then be necessary to sell or to place this expertise in international or national agencies. The presence of members of the standing European Committee for Research on Chronic Toxicity Hazards in the official international committees that were to appear – especially the Committee of Experts for Sanitary Control of Foodstuffs of the Brussels Pact, and the FAO/WHO Joint Expert Committee on Food Additives – ensured the circulation of the general principles and tools that these European experts had started to develop. However, the work of the European committee received a lukewarm reception. Eurotox experts' active participation on these committees failed to provide them with the means to implement the programme that they had designed to protect public health from the dangerous effects of substances added intentionally or unintentionally to foods. Additionally, due to 'reality', as members of the official committees of international agencies on which they served, they also had to give in on a number of other issues that, in the framework of the UICC and Eurotox, they had identified as being the fundamental principles underpinning the production of expertise in the service of public health.

The Necessity of Developing the International Trade of 'Manufactured Foods'

The experts who worked on issues of food additives and contaminants on an international scale in the early 1950s found a demand for their expertise in the UICC and Eurotox, in the Committee of Experts for Sanitary Control of Foodstuffs that depended on certain institutions which laid the foundations for European construction – the Treaty of Brussels between 1948 and 1954, then the Western European Union, between 1954 and 1956, and the Council of Europe from 1956 – as well as in committees dependent on the WHO and the FAO – the WHO Expert Committee on Pesticide Residues, the FAO Panel of Experts on the Use of Pesticides in Agriculture, the JMPR and, above all, the JECFA. Although many of the members of these different committees were scientists on the standing Eurotox Committee, they developed strategies that differed substantially from those of this committee. In fact their strategies were the outcome of a series of instances where they had had to capitulate on certain points, and thus produce tools for expert assessments that would enable industries to carry on developing their activities untrammelled. We will now take a closer look at how this happened, by examining the work carried out within the most visible of these expert committees on food additives and contaminants that depended on international organizations in the late 1950s and early 1960s, the JECFA.

The origin of this committee was a suggestion by the Joint FAO/WHO Expert Committee on Nutrition which, at its first meeting in October 1949, noted that

> food regulations in different countries are often conflicting and contradictory ... The committee draws the attention of FAO and WHO to the importance of this subject to both Organizations, and suggests that its study be included in the programmes of both Organizations.[27]

The process thus initiated resulted, among other things, in the organization of a Joint FAO/WHO conference on Food Additives in September 1955. The report resulting from this conference placed the committee in a dynamic suited to collaboration between the FAO and the WHO, and justified this initiative by arguing for the necessity of establishing international cooperation on the regulation of 'potential health hazards from the inadequately controlled use of food additives', as 'the size of the problem makes it impossible for any individual country to undertake all the investigations needed; lack of facilities and limitation in the number of experts make co-operation on an international basis essential'.[28]

While recognizing that 'such co-operation has already been achieved between certain groups of countries ... and by certain organizations', the report argued that 'unless co-operation is on a wider scale and is more closely co-ordinated, there is a risk of undue duplication of effort'.[29] Thus, it seemed that the FAO and the WHO sought to drive the construction of international expertise on food additives.

The work of the September 1955 conference was based on three reports, one by the FAO, another by the WHO, and the third by the Western European Union, although the latter emphasized mainly legislative aspects of the problem. Chaired by Norman Wright, Chief Scientific Adviser of the UK Ministry of Agriculture, the conference brought together delegates from thirteen countries and four organizations. The US delegation, consisting of four individuals, was the biggest. A third of the delegates were scientists who also participated in other international initiatives concerning the regulation of food additives, and who had been involved in the approach initiated at Bad Godesberg. Truhaut and Druckrey had particular positions insofar as the former represented both France and the UICC, while the latter spoke on behalf of Germany, the DFG and the UICC. One of the decisions taken by this conference was to support the creation of a Joint FAO/WHO Expert Committee on Food Additives whose mission would be 'to formulate general principles governing the use of food additives and subsequently to advise FAO and WHO as to future work in this field'; 'the formulation of suitable uniform methods for evaluating the safety of food additive'; 'appraisal of data sheet on food additives with the object of securing the most suitable form of presentation'.[30] This committee's first meeting was held in 1956. In June 1963 the second Joint FAO/WHO Conference on Food Additives noted that the joint expert committee had convened seven times since its inception, and had produced a set of technical reports.

With regard to the meetings organized in the framework of the UICC and Eurotox, the FAO's and the WHO's approach was characterized above all by economic concerns. While the joint FAO/WHO conference in 1955 stressed the fact that 'the potential health hazards from inadequately controlled use of food additives are worldwide', it also pointed out that:

> Apart from the health aspect, the subject of food additives is one which affects international trade. The greater the measure of agreement on methods of testing, on interpretation of the results, and ultimately on the actual lists of permitted food additives, the greater the advantage to the free movement of manufactured foods in international trade.[31]

The importance of this concern was reflected, for example, in measures aimed at producing 'knowledge of legislative measures in different countries', which was considered as 'no less important' than the 'technical aspect of the problem ... since it should lead to a greater degree of uniformity in the general principles on which the legislation is based'.[32] The FAO and the WHO attracted the attention of the industries concerned very early on. From 1955, the canned food industry sent an observer, Professor Buffa, who was a member of the Standing International Committee on Canned Food – a committee that had initiated procedures to develop international regulation of food additives for its products. The International Committee for Agricultural Industries was also represented. At the second Joint FAO/WHO Conference on Food Additives, in 1963, no fewer than eight international organizations defended the interests of manufactured food industries, whose delegates were present. Even though the protection of public health remained the purpose of the meetings of the Joint FAO/WHO Expert Committee on Food Additives, the necessity to preserve and even to facilitate the production and commercialization of processed foods was a decisive factor in the discussions of this committee, in which 'observers' representing the food industry were always present. This necessity to preserve the agri-food industries profoundly impacted the 'technical' reports – some of which devoted up to a third of their content to justifying the need for food additives – as well as the tools for expert assessments and the regulatory tools that these reports progressively devised.

From 1956 to 1963, this committee produced seven reports. I would like to highlight two of the many interesting elements of these reports: the design of 'Acceptable Daily Intake' (ADI) as a key tool of expertise; and the fact that the idea of banning all carcinogenic substances in food was dropped.

From 'Positive Lists' to 'Acceptable Daily Intake'

Unlike the work carried out by Eurotox, the JECFA was unable to produce positive lists during the period 1956–63, even though many members of the former committee were also on the latter, and one of the missions attributed by the FAO/WHO Joint Conference on Food Additives of 1955 to the Joint Committee was precisely the production of such lists. In 1962 the sixth report of the JECFA, in which the results of a meeting held in 1961 were published, explained this failure to produce positive lists by noting that it was impossible for international organizations to guarantee their application. The report emphasized that:

> It is the common practice in many countries nowadays to work on the basis of 'permitted lists' ... While this may have much to commend it from a national point of view, it is not possible to adopt such a practice internationally, since there is no international law on which permission could be based nor is there any international machinery for enforcement of such legislation.[33]

Another argument was that 'the need for food additives and their possible uses and applications vary so widely from country to country that a single pattern of legislation is not feasible on a world-wide basis'. For the scientific experts on this technical committee, 'it is, of course, clear that groups of countries may agree on joint permitted lists. Such developments are welcome and may represent a valuable contribution to the promotion of international trade.'[34] This position of the scientific experts reflected a tense context at the time, due to the negotiations underway for the creation of the Codex Alimentarius. In 1963, when common ground for agreement had just been found, the second FAO/WHO Joint Conference took a slightly different standpoint on the positive lists, affirming that: 'The Conference considered that the work of the Expert Committee concerning specifications and evaluation of toxicological data on food additives was of basic importance and delegates pointed out that positive lists of additives should emerge from such work.'[35]

While political reasons can explain the absence of the production of positive lists by the JECFA, there were other reasons as well. The JECFA, in a movement that largely transcended this committee – as it also involved the other committees depending on international bodies, and on which the first international experts sat – ended up working on the development of other tools for expert assessments of food additives. One of these tools was 'general principles' on the 'specification of purity' of the different types of food additives and a sound evaluation of their toxicity. But the most important tool developed at the time, which became central in the production of expertise on food additives, pesticide residues and other contaminants, was the ADI.

The function of positive lists was to identify those substances that could be authorized because they presented no danger to public health. At UICC and Eurotox meetings, as well as those of the European committees such as the

JECFA, it was regularly stipulated that 'the amount of an authorized additive used in food should be the minimum necessary to produce the desired effect'. However, it rapidly appeared that this was not sufficient. Although the ingestion of certain doses of specific additives or contaminants seemed not to have noteworthy effects on health, with different doses problems could arise. Not proposing these additives for authorization was considered unrealistic and would therefore have been refused by industry. Likewise, banning pesticides because, over certain limits, residues in or on foods could cause very visible public health problems, seemed impossible. The same question arose from the early 1960s for a whole range of contaminants, notably the residues of 'animal feed adjuncts' and contaminations from 'components of packaging materials which may find their way into food'. Regulations based on threshold limit values therefore seemed to be the best solution. It took time to design a tool, the ADI, which could be used to set these threshold values, called 'tolerance' values. This scientific, technical and political process was carried out by a small group of international experts led by Truhaut, and included Druckrey, Dols, Peter Elias, Alistair C. Frazer and J. Barnes from Britain, the American G. Fitzhugh, the Dutchmen G. van Genderen and G. van Esch, and the 'secretaries' of the JECFA, especially Dr M. G. Allmark (WHO), Dr A. G. van Veen (FAO) and Dr C. Agthe (WHO). It was in a constant feedback loop between European and UN international committees working on food additives and pesticide residues that the committees finally defined what was to become the ADI.

According to Truhaut,[36] and as the various traces left by the creation of the ADI attest, this tool started to take shape in the very early 1950s in discussions on 'the setting of maximal limits for pesticide residues'. These discussions took place in meetings of the Working Party on Pesticide Residues of the Scientific Committee for Health Controls on Food under the Council of Europe. Truhaut 'swiftly came to realize that account had to be taken not only of human consumption of those foods, which varies by country, region and even population group'.[37] He therefore proposed that the experts would need

> to establish a basic concept consisting in the evaluation, on the basis of toxicological data extrapolated to humans, of an acceptable dose for repeated absorption by humans, from which to deduce, having regard to the use of pesticides on certain staple crops and, above all, average rates of consumption of foods derived from them, limits which should not be exceeded.[38]

Truhaut transposed these arguments in JECFA discussions. The ADI as such had not yet been defined, however, and the first technical reports contained only some elements contributing to its definition. In the first report, resulting from a technical meeting in 1956 and published in 1957, one finds for example the idea that 'it may be necessary to set an upper limit on the amount of some additives to be used, either in foods in general or in specific groups of foods'.[39] Truhaut managed to obtain approval for inclusion in the next report, published in 1958 and resulting from a technical meeting in 1957, of an entire section on 'the estimation of the probable safe levels for human use', in which it was explained that:

> A dosage level can be established that causes no demonstrable effect in the animals used ... In the extrapolation of this figure to man, some margin of safety is desirable to allow for any species difference in susceptibility, the numerical differences between the test animals and the human population exposed to the hazard, the greater variety of complicating disease processes in the human population, the difficulty of estimating the human intake, and the possibility of synergistic action among food additives.[40]

Two significant aspects of the design of the ADI appeared in this report. The first was that it was necessary to define 'the standard daily dietary dose', that is, 'the amount of the food additive that might be expected to be consumed by an average adult eating a normal diet as determined from some appropriate dietary survey'. The second was that 'it is inescapable that some arbitrary factor must be applied in order to provide an adequate margin of safety', and that 'the hundred-fold margin of safety' therefore seemed 'adequate'.[41] In Truhaut's opinion, what was to become the ADI was, at the time, still 'somewhat nebulous'.[42]

The concept of ADI seems to have been set in 1961–2. One of the points that appear to have been discussed at length was the unit in which it was to be expressed. Scientifically, it was more accurate to adopt 'a ratio of mg/body surface area'. Yet Truhaut and others argued 'for all practical purposes' that ADI should be expressed in mg/kg body weight. This was the unit that was retained by the various international and national expert committees working on food additives and pesticide residues.[43] The sixth technical report produced by the JECFA, published in 1962 and resulting from a meeting on 5 to 12 June 1961 on 'the evaluation of the toxicity of a number of antimicrobials and antioxidants', started to establish for some of these substances what it defined as 'acceptable intake zones', 'intended to assist those concerned with food additive problems in taking decisions on the use of food additives ... in establishing appropriate tolerance levels'.[44] The term ADI did not yet appear as such in this report, which nevertheless explicitly associated the production of indicators, then called 'acceptable intake zones' by the JECFA, with the production by the competent authorities of norms defining maximum values for additives in certain foods, and that were called 'tolerance'. The first officially published complete definition of ADI is found in a report not on food additives but on pesticide residues. This report, entitled *Principles Governing Consumer Safety in Relation to Pesticide Residues* resulted from a joint meeting of the WHO Expert Committee on Pesticide Residues and the FAO Panel of Experts on the Use of Pesticides in Agriculture, in Rome from 9 to 16 October 1961. The ADI was presented as the 'daily dosage of a chemical which, during an entire lifetime, appears to be without appreciable risk on the basis of all the facts known at the time'. The expression 'without appreciable risk' was 'taken to mean the practical certainty that injury will not result even after a lifetime of exposure'.[45] The ADI was expressed in mg of the chemical, 'as it appears in the food, per kilogram of body weight (mg/kg/day)'.[46]

Apart from the ADI, the report defined several other concepts and tools associated with it, and which served as a basis for the production of standards for threshold concentrations of food additives, or contaminants, especially pesticide residues. In particular, it defined the 'food factor', that is, 'the average fraction of the total diet made up by the food or class of foods under discussion'; and the 'permissible level', which is 'the permissible concentration of a residue in or on a food ... calculated from the acceptable daily intake, the food factor, and the average weight of the consumer'. The permissible level was expressed 'in ppm of the fresh weight of food'. 'Tolerance' that constituted the effective norm, was

> the permitted concentration of a residue in or on a food, derived by taking into account both the range of residues actually remaining when the food is first offered for consumption (following good agricultural practice) and the permissible level. The tolerance is also expressed in ppm. It is never greater than the permissible level for the food in question and is usually smaller.[47]

The definitions retained for pesticide residues were applied to food additives. After evaluating the work already performed by the JECFA, the Second Joint FAO/WHO Conference on Food Additives, in June 1963, added another task to those of the expert committee: '[T]o recommend acceptable maximum daily intake levels.'[48]

Truhaut was presented as the 'obstetrician of the ADI'.[49] He was on all the international expert committees on food additives mandated by the Council of Europe or by the FAO/WHO during the latter half of the 1950s and the early 1960s, at which the ADI was progressively defined. Other, less visible actors appear to have played an important role as well: the 'scientists' working at the FAO and WHO food divisions in charge of issues pertaining to food additives and pesticide residues, who also sat on all the international expert committees and, most importantly, organized those under the FAO/WHO. Dr Allmark and Dr C Agthe are two examples. The ADI, from which 'tolerances', namely threshold doses of food additives, pesticide residues and other contaminants, were defined, had two advantages. First, it was more flexible than systems of banning and authorization, and made acceptable the idea that additives or contaminants which were potentially dangerous to health could be present in foods, while guaranteeing that the quantities in which they were present were not known to cause any major public health problems. Second, its establishment required that many elements which could at any point be revised, discussed and adapted, be taken into account: the methods of toxicological experimentation, the methods of extrapolation from animals to humans, and the calculation of the 'food factor' for example. The many possible points of adjustment made this seemingly highly technical indicator a particularly malleable regulatory tool – far more so than positive lists. For the same additive, pesticide or contaminant, it could be used as a basis for ad hoc regulations that were differentiated according to the economic and political context. The 'food factor' in particular – the average fraction of the total diet made up by the food or class of foods under discussion – could be used to scientifically justify the existence of these differentiated regulations, since it served to construct ADIs and tolerances that took into account the variability of diet.[50] This malleability and adaptability meant that the technical tool for expert assessments, which could be used to guarantee the 'safe use' of food additives and contaminants, could also become a political tool in complex situations where such international organizations as the FAO/WHO or Europe, along with industrial organizations, states and so on, sought to develop certain agricultural sectors, the agri-food industry and trade in agricultural products and processed foods. The fact that the ADI, whose modalities of calculation were repeatedly revised,[51] became one of the pillars of the construction of regulations on food additives and contaminants on an international, European and national scale, is probably due far more to this indicator's capacity to serve the interests of economic development than to its capacity to effectively guarantee the protection of public health – even though it did serve this purpose to some extent.

Giving Up On the Idea of Banning Substances Known to be Carcinogenic

In designing the ADI, the experts and scientists of international organizations endeavoured to take on board both public health concerns and the 'reality' of economic and political imperatives, and thus to produce 'practical tools'. At least superficially, the ADI nevertheless appeared to be based on sound scientific bases – even though the reports published were very careful to point out all the problems posed by the various phases of evaluation of toxicity, and which led them to recognize that 'it is impossible to establish absolute proof of the non-toxicity of a specified use of an additive'.[52] This position was more difficult to maintain in other cases where these experts had to give up the idea of what they considered to be important steps that might guarantee the protection of public health, for the sake of agricultural development and the agri-food industries' interests. They had to agree to many compromises with regard to their conception of what good tools were for expert assessments and the regulation of carcinogenic substances found in food.

At the UICC meetings and the Bad Godesberg meeting, Truhaut, Druckrey and others had argued that 'for substances producing irreversible summative toxic effects ... there is no nontoxic "threshold" dose'.[53] They furthermore considered that those substances whose harmlessness had not been proved in the conditions of their use should not be authorized as food additives. Consequently, substances which were proved beyond any doubt to be carcinogenic in any animal species and by any means of administration whatsoever were not to be authorized.[54] This position is found in the second JECFA report on the *Procedures for The Testing of Intentional Food Additives to Establish Their Safety for Use* published in 1958 and resulting from a meeting in 1957. The JECFA committee, chaired on this occasion by the American R. Blackwell Smith and Truhaut, 'believed' that 'no proved carcinogen should be considered suitable for use as a food additive in any amount'.[55] Adopting this point of view to develop systems of regulation nevertheless posed many problems, of which these experts were well aware. Apart from the fact that some additives used in the food industry were suspected of being carcinogenic, additives often contained impurities resulting from their production – especially heavy metals and arsenic residues which had been recognized as carcinogens for a long time – or were linked to other major public health problems. The experts at the JECFA meeting in 1957 did not, however, discuss this problem any further. They called on the FAO and WHO to 'consider referring the problem of the possible carcinogenic and mutagenic action of food additives to an appropriate group of experts'.[56] This ad hoc group of experts, which produced the fifth report of the JECFA, published in 1961, met in Geneva from 12 to 19 December 1960. It was chaired by Alistair C. Frazer from Britain and G. van Genderen from the Netherlands, and Truhaut was present. Discussions included not only the 'evaluation of the carcinogenic hazards of food additives' but also that of 'food contaminants and substances occurring naturally in food'.[57] The focus was on two main points: the evaluation of the carcinogenicity of the substances under consideration, and the principles on which to base the recognition of the carcinogenicity of substances present in food.

A large part of this report on the 'evaluation of the carcinogenic hazard' was devoted to the definition of good testing procedures. The committee stipulated that 'each individual substance' was to undergo a 'separate assessment of the carcinogenic risk'.[58] As the majority of food additives and contaminants had not been adequately tested, it would therefore be necessary to assess them all. The committee identified two significant limits to the procedures for establishing the carcinogenicity of a substance. The first concerned the fact that 'tests on animals cannot provide irrefutable proof of the safety or carcinogenicity of a substance for the human species'.[59] They however considered it possible 'to formulate practical procedures for the determination of possible carcinogenicity within the limits of our present knowledge'.[60] But in this statement they had to take into account two phenomena that constituted the second limit. Whereas 'many substances ... require study' and 'some permitted lists contain apparently untested substances', the committee 'recognizes that the available facilities and experienced personnel for carrying out such studies are limited and may remain so for an unpredictable time'. As a result, 'proposed tests should be relatively simple and no more time-consuming than necessary to facilitate the testing of as many food additives as possible within a reasonable period of time'.[61] This lack of the required means resulted in a lowering of the demands as regards the characteristics of a 'good' carcinogenicity test and the formulation of a 'minimum safeguard' consisting of 'an investigation of the tumour incidence in a chronic toxicity test as set forth in general terms in the second report of the JECFA'.[62] While the proposed test appeared to have similar characteristics to those described in the UICC and Eurotox reports, the terms used were far less imperative. For example, whereas at the Bad Gotesberg conference it was agreed that 'the tests should be carried out over three generations',[63] the experts on the JECFA committee of 1957, who produced the second report – and some of whom had participated in the Bad Gotesberg conference in 1954 – explained that 'where tumour formation is of primary concern, total life span studies, extending over two generations, in at least one species, have been advocated and may be desirable'.[64] The 1961 report *Evaluation of the Carcinogenic Hazards of Food Additives* made a set of recommendations on how to run tests and interpret their results. Yet the recommendation of a 'minimum safeguard' that had slacker requirements than those presented by these experts in the report, shows their awareness of the unlikelihood of more elaborate tests, which would be more likely to show up problems, being widely adopted as tools for expert assessments. The national or international regulatory authorities in charge of such assessments would not in any case have had the human or material resources to enforce their use.

This forced lowering of the requirements is found in discussions on the principles that were to underpin the regulation of carcinogenic substances present in foods. Two of the committee's statements put it in a difficult position when it came to solving this problem. First, it recognized that 'the presence of carcinogenic substances in food might be a significant factor in the occurrence of what is considered to be spontaneous cancer in man and animals', and second, it acknowledged 'the difficulty of completely excluding at the present time carcinogenic contaminants from food, from processes used in the preparation of food, and from substances coming into contact with food'.[65] It followed that a ban on any substance recognized as carcinogenic in foods became impossible from an economic point of view. The experts on this committee consequently sought to develop what they called a 'practical approach', attesting to multiple difficult compromises, if not impossibilities. The first thing they did was to give up the idea that there was no threshold for carcinogens, and in so doing made it possible to establish tolerance levels for these substances. Thus they explained that, since dose–response relationships have been demonstrated in the case of carcinogenic agents, the reduction of carcinogenic substances in food to the lowest practicable level may be one of the effective measures leading towards cancer prevention. At the same time, on several occasions in the report, they emphasized the fact that 'many factors may influence dose-response in carcinogenesis'[66] and that:

> Carcinogenesis is a complex process ... The uncertainty of the extrapolation of the safe dose to man, and lack of knowledge of the possible summating or potentiating effects of different carcinogens in the total human environment, preclude the establishment of a safe dose at the present time.[67]

They concluded that 'the elimination, or at least reduction to a minimum, of all proved carcinogenic substances in the diet of man and of animals used as human food is a worthwhile objective'.[68] Thus, in this context of official expert assessments, the demand for the ban on carcinogenic substances in foods that certain experts on this committee had vehemently defended, turned into an 'objective' of elimination or 'reduction to a minimum' that was more compatible with the reality of agricultural and industrial practices. This compromise was translated into recommendations aimed at validating regulatory tools as alternatives to the ban on all carcinogenic substances in foods, and which had de facto started to be developed. Thus, the committee acknowledged that 'with carcinogens that occur in the general environment and contaminate food, tolerance levels are set'.[69] It also accepted that:

> In case of secondary contaminants, such as carcinogenic substances extracted from plastics or waxes or resulting from contaminating processes ... specifications are required which will ensure that the contaminating carcinogen is kept to the lowest practicable level.[70]

It confirmed that:

> If a pesticide is found to be carcinogenic, residue may be avoided in food either by banning its use on food crops, by controlled use, or by insistence on the reduction of the residue to the lowest practicable level (that is, the level at which it is no longer detectable by a sufficiently sensitive method).[71]

The conclusion of the report reflected the dilemma facing these experts, and the compromises that they had had to make:

> The committee supports the view that any attempt to establish a safe dose for carcinogenic substances in the human diet at the present time would be unwise. Nevertheless, the Committee points out that permissible levels for some carcinogenic agents are unavoidable, as, for example, with naturally occurring substances or carcinogens of ubiquitous distribution.[72]

Thus, they endorsed the beginning of a movement that sought to justify the presence of carcinogenic substances in foods and more generally in the environment.[73]

This movement impacted on the role that the committee members could claim to play in the production of expertise and of the regulatory tools consequently developed. From this point of view, the JECFA's report on carcinogenic hazards, published in 1961 and resulting from a meeting in 1960, attests to a significant change. In the second JECFA report on the 'procedures for the testing of intentional food additives to establish their safety for use', published in 1958, the experts had stressed:

> It cannot be too strongly emphasized that, just as the conduct of the experiments designed to test safe use is the responsibility of the scientist, so also is the evaluation and interpretation of the data from these experiments. In order that sound decisions can be made regarding the use of food additives, the scientific data that have been assembled must be made available in a form that will allow independent scientists with the appropriate experience to make an adequate assessment of the findings.[74]

In the conclusion they again emphasized that 'scientists working in this field have a responsibility not only to indicate as comprehensively as possible the work that should be undertaken, but also to evaluate evidence upon which decisions on safe use may be made'.[75] The scientists mobilized in expert assessments of food additives, they claimed, had a key part to play in the production of expertise and regulations.

In 1960 the discourse of the JECFA experts at the meeting on 'carcinogenic hazards' was, however, somewhat different. They pointed out that 'it is the responsibility of toxicologists, oncologists and food scientists to evaluate the carcinogenic risk'. They acknowledged that 'the final decision with regard to the action to be taken must depend upon a number of other considerations', that these scientists 'are not the only people needed to evaluate the significance of these other issues', and that 'the decision must be made by a group which should include, in addition, people with other qualifications and experience, such as agricultural scientists and public health authorities'.[76] The carcinogenicity of a substance was not the only element to take into account and, to use the terms of the report, the 'basic issue' became 'to balance risk against benefit from the point of view of the community and the individual consumer'.[77] Thus, the presence of DDT or arsenic in foods that could present dangers for the health of certain consumers was to be weighed up against 'the importance of the substance to the community as a whole'.[78] By adopting this position, the JECFA experts who produced this report took part in the beginning of a dual movement. First, they agreed to a role which was marginal and less important than the one that they had originally claimed to play, in the actual process of production of regulations on substances intentionally or unintentionally added to foods. Second, they also participated in the creation of loopholes which, two decades later at the end of the 1970s and the beginning of the 1980s, led to the introduction of a government through risk in the food additives and contaminants, and more generally chemical substances, sector. The government through risk was to be applied in addition to a government through threshold-setting on which the tools for expert assessments and the regulations developed in the 1950s and 1960s had been based. It was first developed in the nuclear sector and fully embraced the standpoint that was starting to emerge at the beginning of the 1960s in meetings such as that of the JECFA, and in certain scientific publications, that specific groups of people could be the victims of the negative effects of certain technologies that it was considered necessary to use or to develop. With risk it was no longer a matter of guaranteeing the protection of public health, but rather of establishing a 'socially acceptable' level of negative effects on health and the environment.[79]

Conclusion

To conclude, I would like to draw attention to three points.

First, considering expertise in a regulatory framework through the tools that this expertise mobilizes, affords us insight into the influence of certain interest groups or other players on the production of expertise. Admittedly, the study of the production of some of the regulatory tools and tools for assessment developed during the 1950s and the early 1960s at international level to evaluate and avoid dangers to public health in food additives and contaminants, does not enable us to analyse the means through which the actors defending the interests of the 'economy' were able to substantially influence the definition of those tools. However, it does enable us to highlight the fact that the notion of 'reality' was integrated into the very design of these tools, i.e. the ADI, tolerance and the 'evaluation of carcinogenic hazard'. In other words, they incorporated the necessity to avoid imposing strong constraints, which might impede the development of certain industries. Even though the first experts who laid the foundations of international expertise on the dangers to public health of food additives and contaminants had the stated goal of protecting public health, by giving up one objective after the other they contributed to producing systems of regulation in which the role of public health ended up being only relative. Moreover, when our sources inform us of these committees' practices, we see that in taking into account 'reality', the experts went further than they were compelled to by the tools and regulations that they had contributed to defining. A case in point was a discussion on the high levels of aldrin and dieldrin residues found in certain vegetables, during a meeting of the French committee responsible for studying the toxicity of pesticides. In the secrecy of this meeting, Truhaut explained that 'although the acceptable daily intake for these two pesticides was very low, the FAO/WHO committee he belonged to (the JMPR), when setting the tolerance threshold, took account of the reality'. Hence, from the late 1960s this tolerance threshold, set at a higher level than it should have been according to the rules made by the very same JMPR, was already insufficient and was disregarded. Clearly, the existence of tolerance thresholds, set above all to leave sufficient latitude for the industries concerned, did not imply that they were respected.

My second point is that the analysis of the construction of certain tools for assessing and regulating food additives and contaminants shows the high level of fragility of experts and their expertise. Involvement in the construction of international expertise on food additives and contaminants was probably a career opportunity for some of the scientists concerned. Many of the 'pioneers' of the construction of international expertise on toxic chemical substances who, like Truhaut, often went from one sector to another (food additives and contaminants, occupational health, atmospheric pollution, drugs), subsequently became scientific players at international level. Their work in those early international committees probably also enabled them to create spaces for the action and development of the disciplines in which they worked –toxicology in particular. However, this was at the price of what Nancy Langston[80] has called 'a retreat from precaution', and not all of them were able to process this successfully. Druckrey, for example, rapidly withdrew from expertise-related work. When the time came to renew his registration in the pool of experts who could be mobilized for the JECFA or the JMPR, he clearly told the WHO scientists who called on him that he would willingly have accepted, but that as his interest was in understanding the fundamental mechanisms of carcinogenesis and mutagenesis, he was of no use to the technical work of compiling data on substances.[81] Truhaut, on the other hand, engaged fully in the construction of international expertise on toxic chemical substances, and was one of its mainstays until the mid-1980s. However, even though he largely participated in the construction of this expertise, he was certainly not unaware of what it was capable of producing. In 1962, for example, even though he had contributed in the preceding year to writing the JECFA report on 'the evaluation of carcinogenic hazards of food additives' which started to provide data that could justify the presence of carcinogenic substances in foods, he published a long article in Archive of Environmental Health, in which he described what he considered to be a sound evaluation and in which he affirmed that:

> The only prudent attitude, if we wish to avoid using human beings as guinea pigs, is not to allow foreign substances to be present in food until they have been subjected to sufficiently long experiments to establish their harmlessness to man under the conditions of their use.[82]

Third, the international systems of regulation of food additives and contaminants and, more generally, toxic chemical substances, have undergone numerous transformations since the early 1960s, in particular following major crises which highlighted shortcomings with regard to the protection of public health. It seems, however, that certain problems raised by this expertise when it was set up in the mid-1950s and the early 1960s have still not been solved. As criticism levelled at international regulatory agencies' evaluation of the sanitary effects of BPA shows, the fact that the tools mobilized for expert assessments can favour certain interests at the expense of others still appears to be an issue today.

3 FROM THRESHOLD TO RISK: EXPOSURE TO LOW DOSES OF RADIATION AND ITS EFFECTS ON TOXICANTS REGULATION

Soraya Boudia

On 5 August 2011, the day before the Hiroshima Commemoration, *Science* magazine announced the start of a new and highly ambitious study: 2 million people will be monitored for at least thirty years in order to assess the effects of low doses of radioactivity.[1] The study announced by *Science* concerns the entire population of the prefectural region of Fukushima, to the point where Akira Watanabe, vice-chairman of Fukushima University declared 'We're all guinea pigs.'[2] And indeed, for the second time in history, and willingly or not, part of the Japanese population has become an object of study for doctors and scientists. With this new epidemiological cohort, this population has sadly strengthened its position as the leading provider of data on low dose effects of radioactivity. As the United Nations Scientific Committee on the Effects of Atomic Radiation (UNSCEAR) pointed out in its last report, up until this study was announced, the data obtained from the medical monitoring of Hiroshima's and Nagasaki's populations were the main source of knowledge on the effects of exposure to low doses of radioactivity. Because the people living on the land contaminated by the Chernobyl disaster were not systematically monitored over a long period, over coming years the Fukushima data on low dose effects of radioactivity might provide crucial information to complement data gathered from monitoring survivors of the Hiroshima and Nagasaki bombings over recent decades.[3]

Fukushima highlights a problem the importance of which has been constantly growing since the 1970s – that of the controversial effects on health of exposure to low doses of environmental pollutants. This has become a permanent issue in the fields of research and expertise for a wide range of cases: radioactivity, food additives, pesticides, dioxins and endocrine disruptors. It strengthens the idea that we live in a world that is increasingly polluted by chemicals and that a growing number of existing or new pathologies are linked to these contaminations.

This chapter aims to analyse the emergence and development of the 'low-dose' exposure issue, taking the iconic case of nuclear power as its starting point. The nuclear field is central to the conception and management of problems relating to the toxic effects of physical and chemical agents. It was in this field that the question of the effects of exposure to low doses was clearly formulated. It was also within the nuclear field that scientific and regulatory concepts were developed and implemented in order to study and manage the issue. Nuclear power has hence played a determining role in defining the scope of the problem and its modes of existence both within frameworks of expertise and in the public arena. I will begin by describing the two key stages during which the issue of the effects of exposure to radioactivity in low doses came about – the middle of the 1950s and the beginning of the 1970s, both times in the context of major health controversies which led to its definition and to it becoming a permanent feature in the landscape of research and expertise. I will then look at how the development of the low-dose issue caused a major crisis of the dominant paradigm of threshold in the management of toxic substances, a crisis which led to changes in the conception and management of environmental health matters, with the emergence of risk assessment. By retracing the history of the emergence of low-dose effects of radioactivity and by describing the changes this has brought about, this paper is a contribution to a history of successive accommodations and of construction by experts, industrial companies and policymakers of an assumed acceptance of progressive contamination of the environment by toxic substances.

Emergence of a Long-Lasting Controversy

The controversy surrounding the effects of exposure to low doses of radioactivity already has a long history. It came about during debates and mobilizations against atmospheric tests in the 1950s. In the middle of the Cold War, the race to develop ever more powerful nuclear weapons, and the fear that this produced, nourished a major international mobilization of pacifist inspiration. Hundreds of thousands of people declared their opposition to potentially destructive nuclear escalation through demonstrations, petitions, writings and many other types of action.[4] From the mid-1950s, alongside this political movement, another type of mobilization took form; given impetus by scientists, its main theme was the issue of the effects of radioactive fallout on mankind and ecosystem.[5] The spread of radioisotopes from nuclear explosions into the environment caused numerous scientists to fear a significant increase in the number of pathologies. Genetic effects – mutations caused by radiation and its consequences – emerged as a central issue of these controversies. Such effects had been established after the war by American geneticist Hermann Muller, who received a Nobel Prize in 1946 for works carried out on drosophila. The reason that debates focused on genetic risks was that it not only had an immediate effect, but also had effects that were deferred over several generations.[6] It might therefore cause an irreversible alteration in the human species, something which triggered images of the degeneration of the species and of the possible production of Godzilla-like monsters in the Japanese cinema. The risk of radiation-induced cancers was also a major subject of research and debate. There was particular focus on leukaemias caused by the spread of radioisotopes from nuclear explosions into the environment, especially the radioisotopes Strontium 90 and Cesium 137.[7]

Far more than genetic mutations and radiation-induced cancers, several scientists expressed their fear of a generalized contamination of the Earth by radioactivity which would change the nature and scope of the problem of the effects of radioactivity. Although the dangers of radioactivity had been viewed as a standard professional risk which was covered by tight administrative procedures since the period between the two world wars, the problem of the pathogenic effects of radioactivity became a global risk, characterized by a change in spatial and temporal scales of health threats to which humans were exposed, either directly or through environmental contamination.[8] French radiation biologist Antoine Lacassagne, internationally famous for his work on cancers, summarized the new issues raised by the effects of radiation in the following terms:

> Although the harmful effects of ionizing radiation on living organisms taken individually and directly irradiated, are fairly well known, radiation biologists are now facing problems of a new type – because they are collective and even global. Some concern foodstuffs; to what extent can the contamination of earth and water by nuclear fission products affect animals and crops without any risk of making them dangerous to eat? Other problems are of a genetic nature: as from what levels of irradiation do effects on genital glands create a risk to the fertility of species and the quality of their offspring? Massive contamination through nuclear fallout aside, is it possible for the air to be contaminated by fine dust made up of radioactive elements which surreptitiously enter the organism when we breathe? If weather conditions can be modified, as physicists have declared, then will such changes lead to modifications to living conditions on the planet?[9]

It was within the framework of these controversies that the issue of the effects of exposure to low doses of radioactivity emerged, in relation to how to define the maximum levels of exposure to radiation. Determination of these norms was at the heart of work carried out by radiation protection experts between the two world wars. As early as 1928, several physicists and doctors created the International X-rays and Radium Protection Committee in order to make recommendations for protection against the dangers of radioactivity and X-rays.[10] This work was carried out within the framework of what we might call the 'threshold paradigm' of industrial toxicology. As with chemical pollutants, protection against the pathogenic effects of radiation requires the determination of a threshold for doses of exposure, known as 'tolerance doses'.[11] The tolerance dose was thus considered to be the dose which did not cause any significant and visible harmful effects and which therefore presented no risk. The situation was to radically change after the war, with the development of nuclear technologies.

The nuclear field was thus one of the first – if not the first – field in which the 'threshold paradigm', implemented to manage toxic substances, was called into question and debated. At the end of the 1940s, on the basis of the observation that radiation caused genetic mutations whatever the dose may be and that it was impossible to confirm their safety, various American scientists suggested changing the term for exposure limits from 'tolerable dose' to 'permissible dose'. In 1949, the American leading expert in radiation protection, Gioacchino Failla, expressed the meaning of this change:

> The concept of a tolerance dose involves the assumption that if the dose is lower than a certain value – the threshold dose – no injury results. Since it seems well established that there is no threshold dose for the production of gene mutations by radiation, it follows that strictly speaking there is no such thing as a tolerance dose when all possible effects of radiation on the individual and future generations are included. In connection with the protection problem the expression has been used in a more liberal sense, namely, to represent a dose that may be expected to produce only 'tolerable' deleterious effects, if they are produced at all. Since it is desirable to avoid this ambiguity the expression 'permissible dose' is much to be preferred ... In the first place it is well to state explicitly that the concept of a permissible dose envisages the possibility of radiation injury manifestable during the lifetime of the exposed individual or in subsequent generations. However, the probability of the occurrence of such injuries must be so low that the risk would be readily acceptable to the average normal individual. Permissible dose may then be defined as the dose of ionizing radiation that causes no appreciable bodily injury to the average normal individual at any time during his lifetime. As used here 'appreciable bodily injury' means any bodily injury or effect that the average normal person would regard as being objectionable and/or competent medical authorities would regard as being deleterious to the health and well being of the individual.[12]

The scientists' idea was that 'permissible dose' left open the possibility of a pathogenic effect during the course of a person's life, but that the probability of this was low enough for the risk to be 'acceptable' for the average person. Radioprotection was thus constructed on the basis of the notion that any dose of radiation could involve a risk, but that it was a limited risk, which allowed experts to continue with their work of defining the limits for exposure. This was almost a stance of principle, but it involved a hidden contradiction: on the one hand, it was recognized that whatever the level of exposure, dangerous effects were possible; yet on the other hand, they continued to set exposure limits which were supposed to protect people from the dangers of radioactivity.

These debates between radiation protection experts on the management of exposure to radioactivity finally raised questions among groups of scientists who were not directly involved in regulatory activities. If exposure to a dose which was less than or equal to the permissible dose was never without risk, then what was the nature of this risk? This question had not been openly asked by radioprotection experts, but was asked by geneticists, in response to the public announcements made by American authorities who stated that atomic tests were safe because the level of nuclear fallout was in line with the permissible dose. The prestigious California Institute of Technology (Caltech) was one of the main organizations to formulate this question. In July 1955, Georges Beadle,[13] director of the biology division, who was involved in the major joint expertise being run by the National Academy of Science (NAS)[14] on the effects of atomic tests, wrote a one-page memorandum to members of his division, entitled 'Possible direct effects on man of low level exposures to ionizing radiation'.[15] His intention was to encourage works to be devoted directly to observable effects, over one generation, of the effects of exposure to low quantities of ionizing radiation. He stressed the fact that the 'permissible' or 'tolerance' dose supposed that the effects of low doses were insignificant or of very limited scope, to the extent that they were considered to be negligible. Basing his argument on a 1950 study which showed that radiologists had a rate of leukaemia twenty times higher than other doctors, he questioned the validity of what he felt to be mere supposition. In particular, he felt that it was necessary to reconsider the link between exposure to low doses of ionizing radiation and the appearance of leukaemia.

Following this request, in 1957 Edward Lewis, a Caltech biologist,[16] published an article in *Science* which was to be a landmark event in the formulation of the issue of exposure to low doses of radioactivity.[17] In his article he summarized the available data on the effect of radioactivity on the number of cases of leukaemia, based on works on four groups of people exposed to ionizing radioactivity: Hiroshima survivors, young adults in Great Britain treated by radiation therapy for ankylosing spondylitis, children treated by radiation therapy for pathological thymus and a group of American radiologists. He highlighted two conclusions: first, exposure to low doses of ionizing radiation led to an increase in the number of cases of leukaemia, and second, that whatever the dose, there was a linear relationship between the dose of exposure and the effects produced. These scientific results were caught up in the web of public debates on the effects of atomic testing, with anti-test supporters using them to assert that such tests had already caused thousands of cases of leukaemia, both in the United States and internationally. In the opposing camp, experts with links to the Atomic Energy Commission (AEC) criticized the conclusions with a two-pronged argument. The first concerned the heterogeneousness of the study data, in particular the fact that only the radiologists had really been exposed to low doses and that therefore the results were an extrapolation on the basis of high doses of exposure. The second of criticism consisted in highlighting other results, which contradicted those provided in the article, in particular the data from research on animals, which was developing in the field of radiobiology.[18]

The AEC's scientists preferred the hypothesis of the existence of a threshold, which, like toxicological chemistry, postulated the existence of a level of exposure below which the risk was non-existent. Such a hypothesis made it easier to justify atomic tests inasmuch as the radioactivity of the fallout was below the allowed threshold. In parallel to media campaigns to reassure people with regard to radioactive fallout, the AEC funded numerous scientific works on the effects of radioactivity. In addition to scientists with links to the AEC, the majority of experts remained relatively sceptical concerning the real dangers of exposure to low doses as well as to the possibility of defining a single dose–effect model for all types of cancer which would also apply to all individuals. Publicly, these experts adopted a cautious attitude, stressing the lack of available data and calling for additional research, convinced that future results would invalidate the notion of risks from exposure to low doses.

From Technical Norm to Acceptable Risk

Controversies concerning radioactive fallout gradually subsided after the moratorium on atomic tests in 1958, the Cuban missile crisis, and the signing of the Berlin Treaty in 1963 (banning all atmospheric and ocean tests), but without there being any consensus on the effects on health and the environment.[19] In the United States these controversies led to a major institutional reconfiguration by no longer allowing a single institution to have the monopoly for nuclear activities: the AEC was put in charge of the development and use of nuclear energy, whilst the monitoring of these activities – and of related risks in particular – was transferred to a new independent administration. Although up until then the AEC had had a monopoly on all things nuclear, public health authorities managed to get their foot back in the door by helping to create the Federal Radiation Council (FRC) to regulate radioactivity-related health risks. These debates also created pressure for a clarification of conceptions within the international radioprotection system. At an international level, the International Commission on Radiation Protection (ICRP), which had replaced the International X-rays and Radium Protection Committee after the war,[20] formally recognized that there was no level of exposure below which one could confirm that there was no risk, and that any dose was therefore enough to produce a pathogenic effect. As it was impossible to stop or even slow down the nuclear programme, whilst at the same time suggesting that there was an effect whatever the doses of exposure, experts involved in the management of radioactivity risks nevertheless proposed setting a threshold below which exposure to radiation had effects which were either insignificant or repairable. This was a stance based on principle. The controversy surrounding Lewis's works had demonstrated the blurred and relatively impractical nature of the dose model used by radioprotection experts. The ICRP wanted to be unassailable regarding its determination to implement all possible protections against radiation. In order to publicly justify the principles behind recommendations concerning the risks of ionizing radiation, ICRP's experts came up with a series of hypotheses which were akin to agreements: (1) there is a linear dose–effect relationship for the entire range of radiation doses; (2) there is no threshold above which an effect will be produced and below which no effect will be produced; (3) the doses applied to an organ add up as a whole, independently of dose rate or intervals between exposures; (4) the effects of radiation on health are irreversible. In putting these hypotheses forward, ICRP's experts aimed to make a public display of caution, wishing to protect workers and populations from the dangers of radioactivity. As soon as it was formulated, the linear dose–effect relationship with no threshold set off a long series of argument and counter-argument. As soon as they were formulated, these hypotheses led to numerous objections among scientists who felt that they were not based on reliable scientific data. The most controversial aspect of these hypotheses concerned the suggested dose–effect model, i.e. the linear model with no threshold. This model therefore triggered a series of controversies.

The ICRP's approach did not resolve the problem of publicly justifying the proposed norms. As the ICRP recognized that there was no threshold of exposure without some degree of risk and yet at the same time was setting a maximum threshold for exposure, it found itself faced with the difficulty of justifying why and how this threshold had been determined. Its response was to recommend that every possible effort be made to reduce all levels of exposure to radiation and to keep them as low as possible. It looked to lay down general principles justifying the choice of a maximum level of exposure. As far as the experts were concerned, the practical problem was that the maximum threshold would not be unacceptable for the average individual. So the question was how to define a risk 'which is not unacceptable' or a risk 'which is acceptable' to individuals and the population as a whole. The issue of the acceptability of risks became increasingly important in debates on the regulation of health and environmental dangers.

In the work on clarification carried out within the community of radioactivity experts, there was a clear shift towards the idea that norms were social, economic and technical rather than purely technical, and that the problems resided in their social acceptability. From the mid-1960s, the ICRP made public the framework within which radioprotection experts worked, presenting it as principles or a philosophy of work: 'as any exposure may involve some degree of risk, the Commission recommended that any unnecessary exposure be avoided, and that all doses be kept as low as is readily achievable, economic and social consequences being taken into account'.[21] It therefore clearly recognized that, in establishing norms, it did not take into account technical aspects alone, but also took economic and societal issues into consideration. The ICRP publicly formulated what was already being practised behind the closed doors of expert commissions. The approach was to try to find a balance between the different constraints that Louis Bugnard, director of the Institut National d'Hygiène in France, and member of the ICRP's main commission clearly summarized: 'We must on the one hand reach something that is reasonable, protect populations, and on the other hand avoid holding back the development of atomic energy.'[22]

The ICRP's approach demonstrated the way in which these experts, caught between many different ways of thinking, were trying to define a middle position between different points of view. Although their role was to take steps to protect the health of populations, this did not mean that they should hinder or slow down the development of the nuclear industry, despite the numerous uncertainties about the harmlessness of radiation, the effects of radioactive waste on the environment or the possibility of technological accidents.

Arguments, Counter-Arguments and Transformations of a Controversy

In the early 1960s, nuclear power began a new era, with the development of the civil nuclear industry.[23] Existing structures had become more robust in deflecting criticism levelled at them with regard to the implementation of the nuclear programme that had been conceived and developed from the end of the 1950s. But in reality, the period of calm was not to last for nuclear institutions. From the end of the decade, they once again found themselves at the centre of numerous disputes, which severely tested the system of radioactivity risk regulation and the philosophy of protection that the experts had developed over the years. It was in this context that the issue of low doses sprang back into the limelight in spectacular fashion at the end of the 1960s, to take up a permanent position in the landscape of scientific research and expertise.

At the end of 1969, American scientists John Gofman and Arthur Tamplin presented an exhaustive summary of six years of work funded by the AEC, on the links between exposure to low doses of radiation and cancers.[24] The study had been entrusted to Gofman, a renowned scientist who had not taken part in the previous debates on atomic tests. Professor of biology at the University of California, Berkeley and head of the Livemore biomedicine department (one of AEC's largest laboratories), Gofman was famous for his works in nuclear chemistry which he carried out with Glenn Seaborg, with whom he held patents on procedures for separating plutonium from spent fuels and on the capacity to fission uranium 233. He was also known for his pioneering medical work on cholesterol and cardiovascular diseases. He was assisted by a team of a dozen or so people headed by biochemist Tamplin. One of their first tasks was to draw up a list of all radioactive substances and to determine how they are absorbed by plants, animals, water and soil when they are emitted into the atmosphere. This catalogue was to make it possible to determine, for each substance, how it affects each organ of the body. Another area of their work related to radioprotection norms concerned the effects of radioactive effluents the power plants are allowed to eject into the environment.

The results presented in 1969 caused a major uproar due to their origin and content: a well-known scientist, attached to the AEC, publicly announced that the current norms for radioactive pollution, which prohibited nuclear companies from going above the threshold of 0.17 rad, might nevertheless be responsible for an additional 16,000 deaths per year in the United States due to cancer. Shortly afterwards, Gofman and Tamplin increased their estimate to 32,000 additional deaths. They thus estimated that 10 per cent of deaths from cancer were due to the effects of radiation and underlined the urgent need, for the sake of public health, to lower the norms for radioactive effluent emission by a factor of 10.[25] The following month they were questioned as part of congressional and Senate hearings by the Air and Water Pollution Subcommittee, which was preparing new environmental legislation, the Clean Air Act.[26] These hearings offered a broad platform for their conclusions. Their alarming declarations, which they summarized with the phrase 'protection or catastrophe', had a considerable effect on the chairman of the committee, senator Edmund Muskie, the principal architect of American legislation on issues of environmental pollution. He therefore undertook to call upon a series of institutional figures, in particular those in charge of public health services.

Within the space of just a few weeks, a veritable 'Gofman–Tamplin' affair took shape, combining several themes, as will be discussed below. Its momentum owed much to the context at that time, which it helped to strengthen. This context was marked on the one hand by the increasing number of environmental concerns and mobilizations and a change in the scientific and political regime for dealing with them. The second major contextual aspect was the development of nuclear-based electricity production from the end of the 1960s. There had been an important shift since the signing of the Berlin Treaty: the military aspect of nuclear power was disappearing from the public arena, being replaced by promises of civil nuclear power. The AEC, in partnership with private industry, had launched a vast programme of power plant construction for the production of electricity. Whilst being implemented, this programme encountered considerable local reticence, due to fears of reactor accidents or the radioactive contamination of rivers. Gofman and Tamplin's declarations, stating that even when running at a 'normal' rate[27] nuclear production was responsible for thousands of deaths through cancer, not only maintained this reticence, but heightened it and even helped to give structure to opposition.[28] At first, the two scientists refrained from any criticism of the nuclear industry. It was this specific context that allowed the formulation and explicit definition of a 'low-dose' problem, making it a central issue in debates on risks related to the nuclear industry.

An in-depth examination of the arguments exchanged and actions carried out reveals that the definition of the question of low doses went through gradual transformations as the problem filtered into the public domain. At the beginning, by sounding a health alert in relation to the significant increase in the number of cases of cancer, the two scientists defined the issue of low doses as a problem of thresholds and norms. Such a definition had the merit of being simple. It made it possible to call upon the nuclear industry and institutions, as well as upon the organizations responsible for regulating these activities to provide an institutional and political response that was *a priori* a simple one: to lower the norms.

As various journalists pointed out,[29] from the very start Gofman and Tamplin's declarations constituted a frontal attack against the AEC.[30] As in the 1950s, several experts close to the AEC challenged the validity of their results. They made clear not only their disapproval of Gofman and Tamplin's decision to publish results before they had been assessed by their peers, but also their scepticism concerning the pertinence of the issue raised by the two scientists. They based their attitude on the fact that the risks of radiation had been examined, were known and were under control. Far from intimidating Gofman and Tamplin, these reactions strengthened their determination to present their work to a wider range of forums, and in this they received active support from Senator Muskie and lawyer Ralph Nader.

The Secretary of State for health, faced with the development of the controversy in the media and called upon by Senator Muskie, asked the Federal Radiation Council (FRC) for a detailed evaluation of the radioprotection norms that were in force. The FRC then turned to the expert committee of the American Academy of Sciences, which had been in existence since 1964, to advise it on radiation issues. This committee was a more limited version of the one which had been set up in 1955 to carry out a joint expertise on the effects of nuclear fallout. The experts on the committee defended the idea that no new element justified any reconsideration of the issue of thresholds and norms. Publicly, the FRC nevertheless refrained from relaying this stance, be it in the new congressional and Senate hearings in January 1970 or in the media.[31] It insisted on the fact that it was not at the origin of these norms and that their adoption was based entirely on the work of experts in the international radioprotection system. Since the end of the 1950s, the norms governing ionizing radiation were in fact developed on an international scale, as part of a system constituted by the convergence and 'entanglement' of the activities of different committees.[32] The first actor to take its place in this mechanism was the ICRP. Made up of co-opted scientists, it worked to make recommendations from which practical national regulations and codes would be established. The second actor was the United Nations Scientific Committee on the Effects of Atomic Radiation (UNSCEAR), comprised of official representatives from fifteen countries.[33] This committee was responsible for collecting and organizing all information on the levels of ionizing radiation from all sources, natural or otherwise, and to study their possible effects on mankind and the environment. In addition to these committees, there were the inter-governmental agencies: the International Atomic Energy Agency (IAEA) and the European Atomic Energy Community (Euratom).[34] The line of defense chosen by American experts initially consisted in mobilizing the authority of this system and in highlighting the scientific consensus, which prevailed on an international scale. The ICRP, convened by Senator Muskie in relation to the controversy, set out its positions on the issue of low doses. It adopted a more cautious attitude than that of the American experts. It agreed to reconsider the issue against the advice of one of its founding members and principal directors, American Lauriston Taylor.[35] Gofman and Tamplin thus created a split in the apparent consensus that prevailed among international experts. In order to remain in the game, the expert committee of the American Academy of Sciences ultimately had no choice but to agree to consider the problem of the effects of low doses.

Having become the target not only of the AEC but also of the private nuclear industry, Gofman and Tamplin became increasingly anti-nuclear, through a relatively classic process of shifting and radicalizing stances within public controversies.[36] They joined up with other scientists and engineers who were criticizing the AEC's relative haste in building reactors despite the fact that not all safety studies had been carried out. There was a clear shift at this point, from criticism of the norm to criticism of the experts in charge of developing the norms, who were accused of having a pro-nuclear bias. Whilst the controversy raged, Gofman and Tamplin successively published two works, which revealed their core concerns – the social and moral responsibility of scientists. They declared that men of science had a heavy responsibility, as they were henceforth under the obligation to study, in-depth, the implications and consequences of their research. They felt that it was no longer possible for scientists to hide away in their laboratories and devote themselves to 'pure' and 'objective' research. They clearly asserted that researchers must tell the public what they know, what they are doing and what the benefits and risks of their work might be.[37] The nuclear issue was becoming a perfect example of the new type of problems for which society had to define new modalities of decision-making. One famous Washington attorney, Ralph Nader, who specialized in nuclear matters, wrote that scientific and technological choices are political concerns and must be debated as part of the democratic process. Questions such as what benefits the public wish to obtain and what risks they are ready to accept, become vital.[38] The theme of scientists' professional responsibility was taken up by Nader to forge the figure of whistle-blower. Nader ran a campaign, which received considerable media coverage, to encourage scientists, engineers and other professionals to blow the whistle on the destructive or non-ethical practices of the organizations which employed them. He offered legal services to any professionals who encountered difficulties after taking such action and made the Gofman and Tamplin case his first such piece of business.[39] In 1972 his action led to legal protection for whistle-blowers being added to the Water Pollution Control Act (also known as the Clean Water Act).

The problem of 'low doses' henceforth encompassed a series of challenges: criticism of norms, promotion of the idea of scientific responsibility and development of a joint reflection on the modalities of decision-making with regard to techno-scientific issues. By positioning themselves on different 'niches', Gofman and Tamplin had made it possible to combine different micro-mobilizations in relation to common themes, which amplified their claims. They thus considerably broadened their possibilities for action. The problem for their opponents – the AEC and FRC in particular – was that of limiting this process of 'increasing generality' and of confining debates to the scientific and technical question at the root of the controversy.

A Range of Challenges

One of the main traits of the low-dose controversy was the shift back and forth between a precise criticism, of a technical order, relating to the link between exposure and effects, and a generalized criticism of the nuclear industry and decision-making modes within this area.

So for those who wanted a debate on questions of exposure to low doses of radioactivity, there was a permanent back and forth between the determination to stabilize a definition and the opposite desire to try to extend it into a movement of increased generality for which recourse to the public arena was unavoidable. The extent of the public controversy forced both experts and institutional actors to adjust their stances. For some among them, it was important to try to limit the controversy and to work to keep it within closed institutional arenas where negotiations were easier than in public.

The first level of response was the formal recognition of the need to address the problem. For institutions like the AEC and the FRC, it was therefore a case of attempting to set in motion a process to restore the confidence which had been undermined by the controversy and thus to achieve a demobilization, if only partial, of the protagonists who were seeing some of their critical concerns being heard. The FRC asked the Academy of Sciences to carry out a study of low doses by setting up a new committee of experts with a broader range of expertise than those on the committee which had been advising it until then on radiation issues. The Academy of Sciences took great care in setting up the committee on the Biological Effects of Ionizing Radiation (BEIR). This expert committee had trouble getting its work underway, due to rumours of institutional reorganization which would affect the FRC. The latter was dissolved and its activities transferred to the brand new Environmental Protection Agency (EPA) created by the Nixon administration in 1970. At the beginning of the 1970s, the institutional reconfigurations were far-reaching, allowing many actors to see opportunities for major changes in the management of the effects of techno-sciences. Whilst the regulatory landscape looked to have been completely reworked, those in charge of this regulation were nevertheless rarely replaced. After a period of uncertainty, to demonstrate its impartiality and its determination to seriously examine the problem, the Academy mobilized dozens of scientists to examine four aspects of low doses: genetic issues, somatic issues, environmental issues and the impact on biological growth and development. Gofman and Tamplin's works were in wide circulation, but unlike the AEC's experts, they were neither consulted nor convened to hearings.

The BEIR report was made public in 1972.[40] It was the result of work by several dozen scientists from different disciplines and was internationally accepted as the key reference document on the issue of low doses. It contained a partial acceptance of Gofman and Tamplin's assertions, namely those which related to the need to lower exposure thresholds. Indeed, the BEIR report concluded that a dose of 170 mrem caused between 3,000 and 15,000 additional deaths from cancer per year, depending on the dose–effect model used. The report estimated that the most likely figure was 6,000 deaths. This recognition marked a breach in the experts' line of defense. However, it did not do so by condemning past choices, but rather by attempting to demonstrate their full logic, i.e. that the threshold had been determined with a view to finding a balance between the risks and society's need for nuclear development. It concluded that by adapting the norms it was perfectly possible to continue to maintain this balance. These conclusions allowed the EPA to lower the incriminated norm from 170 to 25 mrem, i.e. to a level close to that demanded by Gofman and Tamplin. The BEIR report also recommended using the linear no-threshold dose–effect model, not for its scientific 'accuracy' but as an operational work hypothesis, since any other hypothesis would be impractical inasmuch as it would require individual variations to be taken into account.

The BEIR's work did not just lead to recognition of the low-dose problem, it also redefined the basic issue. BEIR members chose to place strong focus on the numerous uncertainties of the dossier, reminding people that no-threshold effects and the linear nature of the relationship between dose and effect were merely a set of hypotheses developed for regulation purposes. If one considers the BEIR dossier from a scientific point of view, the first main characteristic of the dossier was the omnipresence of uncertainties at different stages of the production of knowledge. The scientific knowledge relating to exposure to low doses on which the BEIR report relied was a mix of results of work in the fields of toxicology and epidemiology. The approaches adopted within this framework might conceal several uncertainties: uncertainties when extrapolating the results of studies on animals to humans, the results of high-dose irradiation when compared to lower doses of irradiation, the latency of pathologies, the difficulty in establishing any single causality, individual variability, and so forth. Furthermore, the selected dose–effect model sufficed to explain the number of additional deaths. In its conclusions, and stressing the numerous uncertainties surrounding the issue of low doses and the impossibility of proving or disproving a certain number of assertions relating to the risks engendered, the BEIR report operated a subtle shift. The fundamental problems essentially became that of the uncertainty regarding activities presenting a degree of risk and that of the decision-making modalities in such situations: scientific expertise alone was not enough to decide whether or not to continue such activities. A decision of that type could not be made simply on the basis of health and scientific arguments; one had to take into consideration a full set of parameters and compromise between different constraints: not just the dangers, but also the benefits offered by a new technology which met a range of needs, especially in terms of energy. The BEIR report stressed the importance of rethinking the risks and called for the development of a general methodology, which would take into account a set of scientific and socio-economic parameters. In this way it opened the road for a new paradigm in the regulation of environmental health and health and safety in the workplace.

In relation to the low-dose issue, the BEIR was now part of a number of expert groups initiating an in-depth examination of how to regulate innovations and the problems which result. This analysis focused on the issue of risk and the technical and societal modalities for its management. In the middle of the 1970s, the theme of 'risk assessment'[41] became the focus of the work of several federal agencies. The EPA, the Food and Drug Administration (FDA) and the National Institutes of Health (NIH), in partnership with the NAS, the National Research Council (NRC) and the National Science Foundation (NSF) set up a joint think tank on a transversal method of assessment and decision-making in relation to activities presenting risks. The main object of its conception and implementation was that of carcinogenic substances. The modalities for the definition and regulation of a limited number of these carcinogenic substances would serve as a framework for regulating other substances. In 1975, the EPA published a guide to 'cancer risk assessment', based on the case of vinyl chloride, and then wrote a more general guide the following year. In 1978, the Office of Science and Technology Policy (OSTP) prepared a general report on carcinogenic substances.[42] Gil Omenn, professor of medicine at the University of

Washington and the Associate Director of the OSTP, summarized the different stances as follows:

> [T]here was quite a struggle between those who insisted on 'zero risk' and those who proposed methods of risk assessment to identify what Lowrance (1976) called 'acceptable risk' and most of the rest of us preferred to call 'negligible risk', realizing that this conclusion was in the eye of the beholder.[43]

In 1979, an American federal inter-agency committee was set up to make recommendations relating to the identification and assessment of chemical carcinogenesis.[44] With the industrial and political stakes being so high, each stage of these works was marked by a series of debates and controversies between environmental associations, unions, industry representatives, scientists and administrations.[45]

Conclusion: Living with Low Doses of Toxicants

Over the last fifty years, the issue of exposure to low doses has been a permanent feature on the landscape of health and environmental risks. The generic expression 'low doses' now applies to exposure to small quantities of pollutants. It is also used to qualify ways of conceiving, defining and studying them in different spaces. Nowadays the theme is broad and the notion is polysemic. First of all, the issue encompasses a series of questions about the effects of low intensity, the latency of pathologies, the variable reactivity of the different sub-groups exposed, the existence of thresholds and the demonstration of cause–effect relationships. Second, the notion takes on different meanings depending on the group of actors mobilizing it, or on the context in which it is used: it can qualify doses which are not 'high doses' and thus doses which do not produce an acute effect; it can also apply to doses where there is no indisputable epidemiological statistical evidence; finally, it can refer to doses received 'under normal circumstances' by populations and professionals which are lower than regulatory thresholds for exposure.

The dominant definition developed by the BEIR remains the one used by the numerous joint expert opinions, opinions which are regularly called upon. The reports that are produced[46] continue to focus on one of the most important elements of the dossier: the omnipresence and persistence over time of the issue of uncertainty – or rather uncertainties. Formulation in terms of uncertainty does not relate to the radiation dossier alone, but to all low-dose effects. Under the uncertainty label we find a certain number of questions that are formidable from a scientific standpoint, such as the multifactoriality of pathologies, and of cancers in particular; the latency before they appear; their causality; the status and nature of the evidence in these pathologies and the type of link between clinical research and laboratory research. Although a majority definition of the problem of low doses and their mode of regulation has emerged and stabilized, it is not accepted by all actors. The best indicator of this is the number of debates and controversies relating to exposure to low doses that involves the various social groups concerned – workers from different sectors developing pathologies related to their professional activity, parents of children enrolled in the same school and showing a suspiciously high level of cancer – or involving experts whose approaches, or convictions, lead to contradictory results in relation to the same exposure.

The emergence of the problem of low doses and its permanent place in the world of research and scientific expertise, as well as in the activities of regulatory agencies, marks an important shift in the ways of conceiving and managing toxic substances and their effects on health and the environment. Recognition that physical and chemical agents – at first radioactivity and food additives, followed by carcinogenic substances and finally endocrine disruptors – have an effect whatever the exposure and dose, has led to the justification of management through norms being severely called into question. This justification was based on the fact that below a certain threshold there were no effects, or only negligible effects. Since World War II, numerous scientific studies have shown that just like numerous chemical substances, radioactivity causes pathogenic effects below exposure thresholds and that these effects are not negligible. On the basis of the justification of management through norms, such substances should have been banned. Yet such a ban was not one of the possible decisions envisaged either by the experts or the policymakers.

The crisis of the paradigm threshold and of management through norms led to reflection on regulatory systems. This reflection led in turn to a search for new ways of legitimizing the use of substances presenting a degree of risk. The health argument could no longer serve as a basis for justifying a decision to ban a toxic substance; other elements needed to be taken into consideration, in particular a substance's economic and social benefits. There was a gradual shift towards management through the assessment of the dangers and a comparison of risks versus benefits. From the end of the 1970s, through a hybridization of the conceptions and practices embodied in the joint institutional work simultaneously carried out in the USA and Great Britain, risk became a central parameter in the governance of numerous health and environmental sectors. The use of risk to manage a set of problems found its most accomplished form in the 'Red Book'.[47] This report by the NRC, published in 1983, clearly defined the paradigm of the analysis, assessment and management of risk, which was to be copied by different international authorities. Risk is thus defined as objectifiable, with the need for knowledge to be separate from political consequences. Risk analysis and management took over from management through norms. With risk, the problem is no longer the activity, its legitimacy or its benefits, but the definition of norms to govern its 'adverse effects'. Potential dangers were not therefore to be banned, but to be tamed. They became a necessary evil, inherent to life in society. Human, political and economic choices at the root of a given activity were rendered invisible, natural and at the very most, necessary. There was a gradual shift in regulatory discourse, from the necessary ban of toxic substances for health reasons to the possibility of living with them. This was an important step on the road to establishing what is known as the toxic world.

4 'LICENSE TO EXPOSE'? OCCUPATIONAL EXPOSURE LIMITS, SCIENTIFIC EXPERTISE AND STATE IN CONTEMPORARY FRANCE

Emmanuel Henry

Introduction

In France the asbestos affair created a wide audience for occupational health issues, which usually receive little public attention.[1] After 1995, with the spotlight on this affair, it appeared that one of the instruments framing the use of asbestos was the setting of Occupational Exposure Limits (OELs). These limits, representing the exposure thresholds not to be exceeded in professional use of a toxin, were set for asbestos at the end of the 1970s. However, although they protected workers from certain pathologies, they failed to protect them from the risk of cancer, as asbestos is a carcinogen without a threshold.[2] OELs have, at worst, been perceived and denounced as licenses (for employers) to expose (workers to risks), and therefore as the cause of delays in the implementation of preventive actions in industry.[3]

OELs, used in many countries to manage certain occupational risks, highlight the contradictions running through all occupational health policies,[4] for while they are supposed to protect workers from risks of exposure to certain toxins, they also have to avoid subjecting industry to excessive constraints.[5] This fundamentally contradictory dimension of occupational health policies is problematical, especially for contemporary democratic states that have the objective of protecting all their citizens from health hazards. Problems of this nature, concerning all occupational health policies, are particularly evident in the case of OELs, for the very fact of setting these values brings to the fore otherwise largely invisible contradictions. This is the case, in particular, for toxins (especially carcinogens) of which the setting of a threshold value implies the maintenance of a considerable danger level for exposed workers.

Notwithstanding these difficulties, a system for managing occupational toxins by means of occupational exposure limits was generalized during the twentieth century, with a steep increase after World War II and then again in the 1970s.[6] The tool was developed primarily in North America, through organizations such as the American Conference of Governmental Industrial Hygienists (ACGIH) or the Occupational Safety and Health Administration (OSHA). These organizations also played a role at international level, contributing to the diffusion of this regulatory tool and to the standardization of the levels defined for various toxins. Yet the uniform generalization of exposure limits was still far from being a reality. The various states were by no means ready to commit themselves, to the same degree or at the same speed, to this type of policy. On the contrary, we find very marked specificities with regard to different orientations of occupational health policies, in turn linked to relations between the state, scientific expertise and the representatives of workers and employers.[7]

This paper shows how OELs were adopted in France with difficulty and opposition, due to the country's very specific approach to occupational health policies. In many countries the contradictions of occupational risk management policies are dealt with fairly well. The US system of adversarial legalism[8] results in highly explicit criteria on which government agencies (mainly the OSHA) base their OELs.[9] The German system built threshold values based on socio-economic criteria for those toxins where it was impossible to propose a value under which there was no danger. This procedure is the one now prevailing in the European Union, since the adoption of the Binding Limit Values for this type of toxin. Indicative Limit Values remain the norm for other toxins where scientists propose thresholds under which they are assumed to have no damaging effects on health.[10]

The situation is very different in France, where the Ministry of Labour has always refused to publicly acknowledge that its policies cause risks to be taken in certain industries. Whereas on the one hand the ministry highlights the protection offered by its policies, on the other hand it plays down and avoids publicizing their discriminatory aspects (which cause major health risks to be taken for certain categories of the population).[11] Distinct from these public discourses, public occupational health policies have been made primarily since the end of World War II in more secluded arenas within which compromises are reached in practice on the public policy instruments to use and the orientations given to those policies.[12] These spheres of negotiation bring together employers and workers' representatives (whether unionized or not), primarily at the firm level, and function with a relatively high level of autonomy. The government authorities seek to rely as much as possible on consensus, in order to avoid having to take a firmer stand. This is how, in practice, the OELs have been used by firms relying on regular publication of international threshold values by the main organization responsible for the prevention of occupational risks in France, the INRS (Institut national de recherche et de sécurité). The use of OELs has thus been kept confidential, to avoid these contradictions becoming too obvious. This has strengthened the traditional characteristics of French occupational health policies, especially confidentiality, a high level of ambiguity between scientific expertise, social consultation and public decision-making, and reliance, in practice, on local compromises between employers and workers' representatives. It was only more recently that these modalities of regular functioning were called into question under the combined influence of a national event, the asbestos crisis, and constraints imposed by the European Union, in a context of weakening trade unionism and social dialogue. This crisis and these constraints have caused the French Ministry of Labour to make the procedures for determining these values increasingly explicit, and to provide for an independent phase of scientific expertise. The result has been gradual changes in policies on setting OELs, which have consequently become more visible. Yet ambiguities and contradictions remain, between mounting pressure on the state to protect workers' health, and constraints linked to the wish to protect industrial interests.

OELs: Taking the Risk of Clarifying Occupational Health Policies

Unlike other industrialized countries, France introduced OELs very progressively and at a late stage. Whereas in many countries exposure limits were established from the end of World War II (the first list was published by the American Conference of Governmental Industrial Hygienists (ACGIH) in 1946), France followed suit only much later. The first mandatory exposure limits were introduced in the mid-1970s and concerned a very limited number of products. In 1980, only four toxins were listed: benzene, asbestos, vinyl chloride and lead.[13] The setting of these exposure limits was moreover often related to external constraints with which France had to comply (e.g. ILO Convention 136 on benzene, and Directive 78/610/CE on vinyl chloride).

The application of exposure limits for occupational risk management was therefore by no means standard practice in France. Even though a mechanism for the adoption of French exposure limits was set up from 1980, its use remained marginal in occupational risk prevention. It was only the accelerated pace of adoption of exposure limits on a European scale, from the 1990s, that prompted France to take up a growing number of OELs at the end of the decade and in the 2000s. This ambiguous attitude of the French Ministry of Labour regarding OELs as an instrument of public action says a great deal about its position, more generally, with regard to occupational health policies.[14] I will use it to analyse two specific characteristics of these policies: the imperative of limited publicity, and their irreducible embeddedness in irreducible rifts between social groups.

OELs: An Instrument Shrouded in Confidentiality

In France, even more so than in many industrialized countries, the international reference for setting OELs was, and still is, the Threshold Limit Value Committee of the American Conference of Governmental Industrial Hygienists (ACGIH) which regularly proposes a revised list of exposure limits.[15] In practice, the French INRS used to disseminate, via its publications, the ACGIH values for the officials in charge of occupational hazard prevention at the national health insurance funds (the Caisses régionales d'assurance maladie – CRAM) and in industry.[16] The labour administration nevertheless sought to set its own values by organizing various working groups from 1980.[17] These groups never had a very precise status and worked directly within the labour relations directorate (Direction des Relations du Travail) which set up the Conseil supérieur de prevention des risques professionnels (CSPRP), a quadripartite authority consisting of representatives of workers' and employers' trade unions and of the state, and qualified personalities.[18] They were at the origin of circulars published between 1982 and 1996, and established 540 indicative exposure limits. Initially these were based on the values of the ACGIH, until more autonomous work was carried out, although always in relation to the values set in other countries.[19]

The lack of willingness to take up OELs and use them extensively for regulatory purposes is evident in several respects. First, apart from the few binding exposure limits mentioned above, all the others were simply guidelines and therefore little more than management tools for use in industry and by the actors of prevention. The circular of 19 July 1982 and subsequent texts referred to 'accepted limits', which left them wide open to interpretation.[20] Secondly, these limits were disseminated in a non-regulatory form, as collections of documents put out by the INRS and the CRAM, on the different existing limits directly applicable in industry,[21] and as circulars that were not legally binding in the relevant area.[22] The fact that these circulars were published not in the government gazette (Journal Officiel) but only in the official newsletters of the ministries in charge of employment and labour, meant that these issues were not brought to the attention of a wide audience.

The state's minimal involvement in determining OELs reflects the typical approach to policymaking on occupational health in France: lack of publicity, soft regulation, pragmatism and reliance on the social partners (firms and unions). On the issue of exposure limits, more specific explanations can, however, be found. OELs, especially when they concern non-threshold hazardous substances (such as carcinogens), force stakeholders to clarify the trade-offs involved. Although adopting an exposure limit is part of a preventive mechanism, it is also a concrete sign that occupational health policies are risk-management policies which tolerate some high risk levels in certain industries or trades. In other words, it ratifies the fact that occupational health policies cause populations' health to be managed differently, depending on the circumstances of their exposure (whether it took place in the workplace or under other circumstances). As I have shown elsewhere[23] and contrary to the practices of other industrialized countries,[24] the management of this discrimination and these inequalities is possible in France because of the semi-confidentiality characterizing it. Hence, contradictions with the principles of protecting the population, prevailing in other sectors of government action, can be kept semi-confidential.[25] As the adoption of an OEL reveals the content of these policies, it entails an inherent risk that the trade-offs made will become known. This emerged when the Ministry of Labour explained why it had delayed including OELs in its prevention arsenal:

> In France, the introduction of OELs into the regulatory system took place relatively late compared to other Western countries such as the US or Germany. At the time France was reluctant to use this tool which did not offer a complete guarantee of the absence of risk. [26]

This highlights the difficulty of accepting the fact that public policies on occupational risk-management generate a high level of risk, rather than the fact of not having tools that offer 'a complete guarantee of the absence of risk'.

This confidentiality surrounding exposure limits is however accompanied by a practice that does allow for the use of this type of instrument in the field. As one of the main promoters of the use of OELs pointed out, until the 1982 circular, the actors of prevention referred to articles of the regulations, which generally stipulated employers' responsibility regarding their employees' health and security:

> The qualitative and general nature of these texts allows for their application in the most diverse situations. Yet it cannot be claimed that they are seen as an industrial risk-management tool by firms. Paradoxically, the prevention services of the CRAM and firms all referred to the famous Threshold Limit Values (TLVs) of the ACGIH that the INRS annual publication had popularised for a long time in France.[27]

This citation clearly shows that the fact of not publicizing the use of OELs did not delegitimize their application in the field. We can even argue the opposite: that the weak visibility of the instrument facilitated its use in the field. This confidentiality made it possible to avoid having to publicly justify its use and explain its health and/or economic consequences.

The Ambiguous Process of Determining OELs

Confidentiality in the determination and application of exposure limits can be related to their process of development and especially to the ways in which scientific expertise is mobilized. As in the case of other occupational health issues, the scientific dimensions were considered during the consultation and decision-making phases, without a separate scientific evaluation phase.[28] This was above all due to the fact that the expert panels called upon by the Ministry of Labour, under cover of producing scientific expertise, took into account far broader parameters, including socio-economic or political considerations in particular. The composition of these panels was particularly fuzzy, with a mixture of different possible criteria of inclusion. The Groupe scientifique pour la surveillance des atmosphères de travail (G2SAT) was the most characteristic example since, in addition to scientists recruited for their knowledge of the problem, its members included professionals working in prevention at the CRAM and the INRS, that is, engineers working in an advisory capacity for firms, as well as experts in industry and scientists with a strong, explicit trade union engagement.[29]

Whereas in the US, Sweden and Germany the creation of administrative agencies or expert committees devoted to essentially scientific problems has gradually led to a dynamic developing around these actors,[30] the French situation is characterized by a far more informal consultation process.[31] The G2SAT, which developed from a committee for consultations between trade unions and employers, seems to be intended primarily to produce consensus between the various stakeholders, while claiming to do so in the name of a scientific framing of problems:

> Most [of the groups] had claims, might I say, with no malevolence, scientific claims. This may be more or less credible and more or less exaggerated. But most of the groups hid behind a scientific label. To my knowledge, none behind economic or union considerations or things like that, even though there were aspects like that, that were mentioned due to the participants' diversity. In principle it was always quote, 'scientific'.[32]

In the case of non-threshold hazardous substances – that is, those for which it is impossible to determine a value below which there would be no risk – the construction of OELs is impossible from a strictly sanitary point of view. To effectively protect workers, banning the substance is the only option.[33] Proposing an OEL therefore takes as a starting point the acceptance of a risk for workers, with the aim of maintaining economic activity in the sector concerned. This implies that a standpoint has been taken on social and economic matters. In some countries the process of consensus-building between the different actors has been scientifically formalized with tools developed by economists used to evaluate various prevention scenarios.[34] This is what the UK opted for with the establishment of a system of Workplace Exposure Limits (WELs) in 2005.[35] In France, however, despite the work of various researchers in favour of this approach,[36] the economic evaluation of public decisions has barely evolved.[37] The aim has been above all to obtain the approval of the various stakeholders and to build consensus between them, notably with the economic actors playing a major role. In a context marked by unequal capacities to mobilize expertise, between employers and the other parties – be it workers' trade unions or the

Ministry of Labour – this has amounted de facto to granting a preponderant role to the actors from industry.

> On manganese ... initially the representative from industry was in agreement. He had participated in the discussions and we'd reached consensus on the basis of scientific studies. And then at the next meeting, he must have been given a dressing-down by his bosses, he ended up saying that the interpretation of these studies wasn't the same. It had to be questioned and analysed in more depth, etc. And eventually we couldn't pass the proposals that we'd formulated together. In fact, that's partly why this working group was dissolved. The Ministry of Labour must have felt somewhat uncomfortable and started to come under too much pressure from industry.[38]

> The industries reacted strongly when we started hassling them with mandatory limits; one after the other, the wood industry, and so on ... In the case of acrylamide, there was a maximum of pressure on the minister. Regarding ceramic fibres ... we had Saint-Gobain there and everything ... Regarding crystalline silica, there we were virtually threatened. Everyone: me, the deputy director, the director. At the time, I remember, we had a meeting that we all attended, it was a veiled threat, barely veiled ... that we were bad, that we understood nothing. At one stage, on cements ... I still remember the CEO threatened me, the head of all the French cement manufacturers. He was a former director in the central administration and he said to me: 'I've dealt with more difficult people than you', or something like that ... What a time! ... It wasn't always easy, those negotiations.[39]

At each stage of the process the problem therefore remained fundamentally split between contradictory representations and interests. In this context, the modalities of the construction of OELs in France constituted an exercise in which scientific expertise and social dialogue functioned in parallel, without any distinction between the different stages and without it being possible to find a consensus on certain aspects of the problem. This incorporation of scientific expertise into consultations was nevertheless called into question from the mid-1990s.

Increasingly Autonomous Expertise and Relative Reassertion of the State's Role

These processes of determining exposure limits have been changing rapidly since the mid-1990s for several reasons: the consequences of the asbestos affair (leading to industry being sidelined in the decision-making process); the influence of the European Union; and the redefinition of the role of expertise in public health policies, notably with the creation of sanitary agencies. These changes are taking place in a context characterized by the decline of the trade unions and of the dynamics of social dialogue. As a result, the Labour Ministry and its administration are compelled to play a more active part in defining occupational health policy. Although these factors have different origins, they have produced similar effects, and it is difficult to distinguish their respective roles in the changes under way. They have led to the establishment of what appears to be a stricter separation between the scientific expertise phase and the decision-making phase, which can also be analysed as a shifting of the balance of power between the different actors involved in these policies. The state and public agencies thus appear to play an increasingly central part in this policymaking, yet without going so far as to take the risk of entering into conflict with the economic actors concerned.

Change under Constraint

The asbestos affair led to many challenges to the regular functioning of the Ministry of Labour and the actors collaborating with it.[40] The most virulent criticism concerned the excessive weight of industrial interests in the decision-making process. The sidelining of industry took place primarily by way of the independence granted to scientific investigations on occupational health issues. For the government administration, the separation between identifying scientific risks and managing them was a way of breaking away from industrial actors, primarily by acquiring its own sources of expertise.[41] This led to in-depth revision of the procedures for determining OELs, especially the articulation between scientific expertise and political decision-making.

In 1996 the G2SAT ceased its activities. It was then decided to form two expert committees, each responsible for investigating one of the aspects of the problems related to OELs. The first committee, called 'health effects', consisted of toxicologists, epidemiologists, occupational health doctors and specialists in that field. It had to analyse the scientific data to determine the levels of risks associated with exposure to certain hazardous substances. The second committee, 'metrology-exposure', consisted of engineers, chemists and specialists in metrology. It was responsible for analysing real capacities to measure exposure levels, and for informing the authorities of levels of exposure in industry.[42] These committees operated from 2001 to 2005, when their preliminary results were transferred to the national agency for occupational and environmental safety, the Agence française de sécurité sanitaire de l'environnement et du travail (AFSSET)[43] under the Plan Santé Travail 2005–2009 (a five-year national health plan adopted in 2005).

> Afterwards, things started to be separated, which is very logical in theory but difficult in practice. The idea was that expertise on toxicology is one thing, and that it's just toxicologists among themselves and not toxicologists from industry; we don't want them. We're independent. That was after asbestos, 1996, and so on. We don't want anyone from industry in the group any more. We separate things. We really had the CPA[44] syndrome, the asbestos standing committee really marked us. We don't want anyone from industry, so we do toxicology, pure independent experts and others. Now it's been transferred to the AFSSET, clearly, that's logical. And another group is looking at metrology. Feasibility is considered only from the point of view of metrol-

ogy. The reality in the field is no longer considered at all. So the exposure limit is set on the basis of these toxicological data, primarily, and the possibility that we have to measure them or not.[45]

These developments were linked to rapid changes in European regulations and modes of functioning. Initially, the Commission adopted two directives (91/322/CE and 96/94/CEE), setting a number of indicative exposure limits. The creation of the Scientific Committee for Occupational Exposure Limits (SCOEL) in 1995 – which developed from a more informal group, the Scientific Expert Group (SEG) that existed from 1990 to 1995 – was rapidly accompanied by a more formalized process of determining exposure limits. Directive 98/24/CE defines two types of exposure limits: indicative limits and binding limits. The former are reserved for hazardous substances for which a threshold of innocuousness can be determined; the latter, very limited in number (lead, since 1998; asbestos since 2003; benzene, vinyl chloride monomer and hard wood dusts, since 2004), correspond to those substances for which no risk level is considered harmless, and which require a phase of deliberation to determine an exposure level acceptable to all the stakeholders.[46] The adoption of the 1998 directive and the promulgation of various exposure limits at European level led the member states to alter their approach to determining exposure limits, in order to comply with European regulations. In the French case several changes were made.

First, exposure limits disseminated by circular only were granted the status of regulations. This meant that indicative exposure limits would systematically have to be endorsed by orders from the ministry, and binding exposure limits would be confirmed by government decrees. Second, the decision to set up strictly scientific committees was taken, due to the following constraint: when an exposure limit different from the one existing at European level was adopted, the member state henceforth had to supply the scientific facts on which it was based. This constraint was a factor leading to the creation of committees with strictly scientific competencies. Note, however, that at no stage did the systematization of the process of determining exposure limits on a European scale, and the creation of a European expert group, lead to suggestions that most of the scientific expertise work be transferred to the European level.

The Implications of Changes in the Rules of Determining OELs

This series of changes culminated in the transfer of scientific expertise to the AFSSET, under the national occupational health plan (Plan Santé Travail) adopted in 2005. In the same year, expertise in the field of occupational health was added to the remit of the AFSSE (Agence Française de Sécurité Sanitaire de l'Environnement), which thus became the AFSSET – with a T for 'Travail' (labour). A specialized expert committee, the Comité d'experts spécialisés (CES), was created in 2007 'with a view to establish occupational exposure limit values for chemical agents'. It is still too early to analyse all the consequences of this decision. At the time of writing this paper, only seven reports have been published by the AFSSET, two of which are on non-threshold carcinogens (asbestos and refractory ceramic fibres (RCF), as formaldehyde is considered as a threshold carcinogen) and no decision has yet been taken by the labour administration (DGT) on the basis of these reports and opinions. We can nevertheless make some assumptions of the developments under way. The example of refractory ceramic fibres is significant in this respect, as the first non-threshold carcinogen examined by the AFSSET. This issue emerged in parallel with that of asbestos (whose ceramic fibres were used as a substitute in certain applications) and was adopted for investigation by the French administration in the mid-1990s. Yet the decree setting an exposure limit for this material was not issued until 26 October 2007. It set a limit of 0.1 f/cm^3. This was preceded by a transition period, ending on 30 June 2009 and authorizing a value of 0.5 f/cm^3.

The level of 0.1 f/cm^3 is derived from the OELs for asbestos. It was chosen to get round the report of the 'health effects' working group of the Ministry of Labour, which in 2005 proposed a level of 0.03 f/cm^3 that the ministry considered too low for industry.[47] Yet this choice of setting the limit at 0.1 ran counter to the ministry's claim that in this case 'only sanitary criteria were considered in determining the binding exposure limit; criteria of technical feasibility were taken into account only for setting application'.[48] This amendment of the OEL by politico-administrative actors was met with scepticism among the AFSSET officials when they examined the file.

> For refractory ceramic fibres, the expert group that had worked for the ministry had come down for OEL, I think, to an exposure of 0.03 fibres per millilitre, and this was raised, after discussions at the Conseil Supérieur, the MEDEF, workers' unions, etc., to 0.1, even with a transition period of 0.5 fibres per millilitre. ... So there can be a very wide gap. ... In general, on most of its opinions and reports, the Agency's opinions are respected. But on the OEL, at this stage, as it's just the start, we don't know how it's going to go.[49]

As regards non-threshold carcinogens, the AFSSET had difficulty in deciding whether it should explicitly propose an OEL to the ministry – thus implying the explicit maintenance of effects on the exposed population's health – or simply assess the risk levels induced by different OEL proposals. This was the subject of numerous deliberations since, initially, the AFSSET had been asked to propose exposure limits in all cases, including for non-threshold carcinogens – as in the case of RCF. The deliberations that took place within the expert committee highlight the internal divisions that emerged around this issue, thus revealing these experts' representations of OELs and their uses. It shows that determining an exposure limit leads to debates that cannot possibly be limited to only that. On the contrary, they cause OELs to be addressed from the more general point of view of the prevention of occupational hazards.

> On ... non-threshold carcinogens, where we know that there is no threshold below which there is no effect, but where the CES has nevertheless advised to try to determine limits, the question is: in this case, do we set a threshold or not? Given that behind that the implication is: 'Ok, we say there's no threshold, so we don't set anything, we evaluate risks and we try ...' Here too, the debate ... it's that we, we're on the risk-evaluation part and not risk-management. The experts don't see themselves ... as being authorised to take the decision to say: 'we set the risk at 10^{-3}, 10^{-4}, 10^{-5}, 10^{-6}'. So we're sure: we're going to make proposals with different risk levels and, opposite them, different exposure levels. It's up to the administrative authority then to take the decision, to organize the debate, so that it can be set. ... And another opinion shared by the experts is: 'Yes but we know very well there too that if we don't set an exposure limit ...' Because the exposure limit, it's not only an authorization to expose, in France it's: there are other measures that have to be taken on carcinogens, first we substitute ... It's one tool among others to limit exposure, and so at any time it does amount to setting a high standard. And some say: 'but if we don't set anything at all, there's the danger that there's no high standard set'. And so, there too, authorizing exposure to levels ... And it's true that it's not easy.[50]

Following these internal debates, the option was taken by the CES (expert committee) on OELs at the AFSSET not to explicitly propose exposure limits for non-threshold carcinogens, and instead to indicate, where possible, the exposure limits corresponding to the risk levels 10^{-4}, 10^{-5} and 10^{-6}. This opinion was accompanied by a critical analysis of existing OELs. In the case of RCF, the text was the following:

> The value of the OEL-8h (0.1f/ml), that came into force on 30 June 2009, corresponds, in the maximalist scenario of the above model (40 years of exposure 8h/d, 5d/week and 50 weeks/year for 40 years), to a lung cancer excess risk at the age of 70, of 5.10^{-4} (indicating the probability of 5 additional cases of lung cancer for 10,000 workers exposed for 40 years to a concentration of refractory ceramic fibres of 0.1 fibre per millilitre. ... The value of the OELs for 8h of RCF currently set in France at 0.1 f/cm^3 is the lowest in Europe. It is a significant step forwards in the reduction of the risk of exposure to this potential carcinogen for humans (classified 2B by the IARC). However, it must be considered as a step only, and be accompanied by a huge effort at substitution.[51]

Apart from the value effectively adopted, the AFSSET report implicitly evaluated existing OELs by giving an opinion (albeit very indirectly) on the state of the regulations. In the case of RCF, the AFSSET's analysis seemed essentially positive and endorsed the decision. Even if other prevention measures provided for by French and European regulations (above all substitution) were mentioned, the fact of noting that this was the lowest value in Europe and 'a significant step forwards' can be analysed as a point of merit. This judgement is based on a comparison of existing values at international level. Due to the high level of international integration of the scientific community, decisions on OELs depend on those in force in other countries worldwide. For instance, even if it is justified by scientific arguments, the fact that the value of 0.03 fibres/cm^3 proposed by the ministry's working group was refused, probably also relates to the fact that this value is much lower than those existing internationally (the values of 0.2 and 0.5 respectively are cited for the ACGIH and the NIOSH–US National Institute for Occupational Safety and Health).

> The Direction Générale du Travail (DGT) asked the CES on OELs to assess a report written in 2005 by a working group at the DRT, to be used to draw up a proposal on OEL-8h. It proposed an OEL-8h at 0.03 fibres/cm^3, taking into account the levels of exposure associated with pleural plaques in the American cohort (Lockey et al., 1996). However, the CES for OELs decided not to approve this proposal, for several reasons:
>
> – The CES for OELs deemed the evaluation of exposures in the American cohort to be unreliable from a metrological point of view;
>
> – The choice of an exposure level (Lockey et al., 1996; Lockey et al., 2002) in the calculations of the OELs, as well as the taking into account of the low value of the confidence interval in the calculations, are not justified.[52]

How is the French Ministry of Labour going to position itself in relation to the agency's reports? On those hazardous substances for which a threshold has been determined, it will probably be fairly difficult for the ministry's labour administration to justify a different threshold. This could trim its room to manoeuvre down to a matter of timing, that is, the date of official application of the texts. In the case of non-threshold hazardous substances, the AFSSET's opinions will provide another constraint, in addition to all the existing values at international level, and will limit the range of options available to the DGT officials.

> This text [proposing an OEL] was then submitted again to the Conseil supérieur de la prévention des risques professionnels, with discussions where normally the debate is really focused on what had been said: 'We don't change the level of the exposure limits, we set only the application dates'. This explains that for certain values, you've got dates that have been set, like on benzene, there were dates that were applied. Same for refractory ceramic fibres.[53]

We would be wrong to consider these intervals granted to industry before the application of an OEL as insignificant. On the contrary, they reveal a key issue in the determination of OELs. In the case of RCFs, whereas fears as to their carcinogenic nature appeared during the 1990s, an exposure limit was not adopted until October 2007, with an exemption until 30 June 2009. Yet in a meeting of the CSPRP committee on chemical, biological and physical atmospheric hazards, mention was made of the publication of the decree for 2006 with an exemption until 31 December 2007, the date that the firms in the industry, consulted by the ministry, had reportedly considered to be realistic. On two occasions the representative of the confederation of French industry, the Mouvement des Entreprises de France (MEDEF), had intervened to challenge this decision, by proposing to wait for the results of the AFSSET's research, and then by asking whether the government administration had carefully assessed the impact on industry.[54]

The implications of the timing of the entry into application of OELs are crucial when there are stringent constraints weighing on decisions. Because it is increasingly difficult to apply decisions that are not consistent with values decided at international level, and because it is highly likely that a decision close to those values will eventually be ratified, the only option for those actors who want to avoid a new regulation is to try to delay its entry into application. In other words, the timing of the entry into the force of an exposure limit is not just an appendage to this regulation; it is a constituent element. Months, even years, without regulations enable industries to anticipate expected decisions, either by preparing firms for their future application, or by delocalizing to other countries with less stringent regulations. Given the speed at which industrial production processes evolve, and the international competition between firms in these processes, winning time can considerably alter the way in which a new OEL is handled. This question of timing is quite the opposite for the AFSSET, which has to publish its first reports quickly to establish its credibility on the institutional scene of sanitary security agencies.

This contrast between the interests of the agency and those of the employers' representatives shows that current changes in France transcend issues related to scientific approaches to determining OELs. They concern, more globally, the actors who control their production process and above all their implementation. As regards the process of determining OELs, the appearance of the AFSSET on the scene has changed power relations between the various actors involved. It can be analysed as, overall, a strengthening of the weight of the public actors, even if the AFSSET's claim of some autonomy has sometimes led it to take a stance that differs from that of the labour administration. It also leads to the development of new relations with the public sphere, primarily by making all its reports and opinions public, and thus making it difficult to alter any of the values determined and published by it.[55]

Conclusion

Faced with a problem on which various social groups have been structurally opposed for a long time, the search for consensus – even when it is carried out in the name of scientific values – is by no means fruitful every time. The setting of OELs and their implementation in France clearly highlights this impossibility of disembedding the scientific dimension of the production of these norms from the power struggles between groups. The visible rise of scientific expertise in the process corresponds above all to the growing demands by public-sector actors for a bigger share in the decision-making process, in relation to the other actors involved. By redefining the setting of OELs as a scientific issue, the French Ministry of Labour hoped to strengthen its control over the process. However, the state's and the Ministry of Labour's ambiguity are noteworthy, since their new attitude was adopted under constraint, almost exclusively due to outside pressure, and with due regard to the interests of the main industrial firms concerned directly by the results of these policies. More profoundly, the growing role of the state can also be interpreted as a role occupied by default: it is because legitimization of decisions through social dialogue is waning, that alternative modalities of legitimization are gradually being sought in scientific expertise. One also has to remember that the French Ministry of Labour remains a weak sector from an administrative point of view, even by comparison with public health, which also tends to be weak in France. Thus, even though significant changes are witnessed, their capacities to bring about lasting transformations in power relations and forms of public action in the field of occupational health in France must not be overestimated.

More generally, a study of the French case shows how a harmonization or convergence of public policies can be observed from an international point of view. In this case, we can by no means talk about international standards being imposed at national level. Whereas the influence of the ACGIH and later of the European Union is unquestionable, the standards put out by these authorities are not applied as such, independently of the national context. On the contrary, it is for reasons peculiar to national configurations that public policy changes can be witnessed. If OELs are gradually being appropriated as legitimate instruments of public intervention, it is because they enable the public actors to reassert themselves: by promoting an instrument widely used by industrial actors, they avoid direct conflict with them. Likewise, if a growing and increasingly autonomous place is granted to the production of scientific expertise, it is partly because it makes it possible to align occupational health policies with other public health policies. Thus, far from revealing a single definition of threshold values or occupational health, the different national policies on these issues require an analysis that takes into account the national forms of articulation of relations between the state and social groups.

5 CHEMICAL INFRASTRUCTURES OF THE ST CLAIR RIVER

Michelle Murphy

Dead life falls to the bottom of the seabed, settles in the mud. Sand and silt slowly gather above. Time passes, sediment presses, seas recede. Under pressure the once live substance waits. Millennia pass, then, in 1858, oil that has lain in the archive of sediment for so long is pulled up into activity by North America's first commercial oil well in Oil Springs, Ontario, Canada. Oil gushes higher than the treetops, slathering the workers and the land, turning the Black Creek black, running into the Sydenham River which then empties nearby into the St Clair River.

The St Clair River forms a natural passage between wide Lake Huron and shallow Lake St Clair. Water then continues to flow south as the short Detroit River, which then empties into Lake Erie. Within the Great Lakes region that straddles Canada and the United States, the St Clair and Detroit Rivers together cut a border between the United States on the west bank and Canada on the east. Water flows from mostly white Sarnia at the top of the St Clair River down towards the ruins of contemporary majority black Detroit. Today, the St Clair is a deep water shipping channel connected to inland shipping laneways leading out to the Atlantic Ocean. The river is crossed by bridges that carry railroads and highways that then go on to traverse continental North America. What began with one oil well and a short pipeline to the nearby town of Petrolia, is today Canada's Chemical Valley, where 40 per cent of the country's petro-chemical processing is accomplished.[1]

Chemical Valley – which sits just below the city of Sarnia, Ontario and across the river from Port Huron, Michigan – first expanded beyond oil with a polymer rubber processing plant. Dow Chemical opened the plant in 1942 as part of the Canadian war effort. Gracing the back of the 1971 Canadian ten dollar bill, the plant etched in purple ink is a symbolic origin point of a national industrial history. Today this polymer plant is joined by over sixty other refining plants. Oil now pulses through Enbridge's transcontinental pipeline for 2,306 km from the Alberta Tar Sands to be processed in Chemical Valley and distributed across North America and beyond. A web of pipelines leads down to the United States, and interconnect local plants. Chemicals made in one refinery are pumped into another in a matrix of interconnected industrial frenzy. Some twenty-eight pipelines connect the US and Canada beneath the St Clair River. The river which runs past Chemical Valley has become a dense hub of industrial activities, mobilities and histories, a crucial crossroads in the petrochemical driven economies of North America. Called 'The Corridor', the waterway is the birthplace of both the commercial oil well and the Ford Motor Company. Petroleum was reproduced as gasoline to feed cars, cars themselves were made, steel forged, rubber refined, but so too was styrene, chlorine and perchloroethylene (for dry cleaning), as well as plastics that have become the consumer props of everyday twentieth century habits. Canals were dug, pipes buried, power grids laid, towns created, coal burned, jobs found and lost. Petro-chemicals, through Chemical Valley, have brought new things to life.

Starting with the petro-chemical history on the St Clair River, this paper is an attempt to think about infrastructures, forms of life and time in a site inundated with industrial chemicals.

Chemical Infrastructure

What counts as an infrastructure? Concrete assemblages of factories, pipes, jobs, machines, roads and utilities, is a probable response. Sarnia-Lambton County, in which Chemical Valley is located, is celebrated for its industrial infrastructures. Its bridges built especially for transporting hazardous waste, its pipelines, smoke stacks and its eight power generators together form the substrate for a 'vast network of petrochemical and refining processes' that quintessentially qualify as an infrastructure.[2] The science and technology studies scholar Susan Leigh Star has rethought the common understanding of infrastructure as a physical supporting system and instead has reframed infrastructure not as a thing, but as a multidimensional set of relational properties that become an 'ecology of infrastructure'.[3] They are arrangements that structure relationships that emerge from human and non-human activity, which then go on to shape life again. Infrastructures are recursive, their temporal reach formed of many looping effects. Infrastructures are spatial arrangements of relationships that draw humans, things, words and nonhumans into patterned conjunctures. In other words, for Star infrastructures are built ecologies. Exactly what activities, relationships and temporal reaches make up a built ecology is not necessarily obvious. Infrastructures, as Star argues, are typically transparent to us, in that we don't particularly notice them unless something goes wrong. What is included in an infrastructure is an open question; what relationships are seen as making up a built ecology is contestable. Not all aspects of infrastructures are celebrated in the glossy brochures of business groups and corporate partnerships. Some relationships are harder to track, harder to perceive and harder to arrange. Thus, this paper asks questions about the built ecology of this petro-chemical intensive corridor. What sedimented relationships, activities and material dispositions structure life along the St Clair?

The infrastructures along the St Clair River are made up of more than visible pipes, roads and refineries. They include what I will call *chemical infrastructures* that materially and unevenly shape human and non-human life in time and space.[4] By chemical infrastructures, I mean the spatial and temporal distributions of industrially produced chemicals as they are produced, consumed, become mobile in the atmosphere, settle into landscapes, travel in waterways, leach from commodities, are regulated (or not) by states, monitored by experts, engineered by industries, absorbed by bodies, metabolized physiologically, bio-accumulate in food changes, break down over time or persist. Chemical infrastructures are regulated and ignored, studied and yet filled with uncertainties. Many different disciplines and communities of experts make knowledge about chemical infrastructures, but this happens piecemeal – some studying fish, others engineering smokestacks, some tabulating emissions, others diagnosing illnesses. Chemical infrastructures, importantly, are spatially and temporally extensive. They are translocal, connecting moments of production and consumption, moving across borders, and traversing scales of life. They are temporally distributed, as some chemicals break down quickly, and others persist and are present for the *longue durée*, some causing immediate responses in organisms, others only provoking effects that take generations to see, working a slow injury on ecologies or organisms, or even planetary atmospheres. With the term chemical infrastructures, I am naming the pathways and processes of industrial chemicals as it structures the lives and times of both human and non-human.

Chemical infrastructures have been built into the landscape of the St Clair River for over a century. The forms of life possible here, for humans and non-humans, are deeply shaped by the manufacture, regulation and dispersion of industrial chemicals: industrial chemicals are linked to creating jobs, widening rivers, changing ecosystems and altering physiologies. Twenty-five years ago Donna Haraway asked 'Why should our bodies end at our skin?', offering the 'material-semiotic figure' of the cyborg as an ontological politics for attending to the ways living-being was already constituted via technoscience.[5] At that moment, feminists were resisting a politics that posited bodies as natural entities, and instead insisting that any version of 'nature' or 'biology' at the end of the twentieth century was *already* conditioned by technoscience. To this we might add, that any reference to life in the twenty-first century is already conditioned by the chemical distributions of industrialism. Thus, in a similar spirit, one might pose the question, what is the place of industrial chemicals in the built ecologies that structure and constitute life? What figuration of infrastructure does environmental politics demand that we notice?

The notion of chemical infrastructures is meant to highlight the *uneven spatial and temporal distributions* of industrial chemicals and their diverse effects on life. What aspects of life are fostered and what aspects are abandoned in a chemical infrastructure? What is reproduced through chemical infrastructures and what averted and abandoned? What purposefully persists and what is unintentionally altered? Chemical infrastructures are importantly not just about the present, but also about the past; they include sedimented relationships, what happened that is not known, what persists and lurks. Time in chemical infrastructures often works through remainders, buried relationships or creeping accumulations. The temporal pace of chemical infrastructures is often slow or delayed, what David Nixon calls 'slow violence'.[6] Knowledge of past chemical events can be difficult to recover. At the same time, we may still be waiting for the effects from the past to show themselves.

It is this temporal aspect of chemical infrastructures – their slowness, their persistence, their creeping accumulation, their latency – that this portrait of the St Clair River emphasizes. Latency is a synonym of lag. Technically, it is the period of time between a stimulus and a response, the gap between one event and another. In medicine, latency time is similar to an incubation period. Latency time is the lag between infection and infectiousness. Or it is the wait between chemical exposure and symptom. To be latent is to be dormant, a potential not yet manifest. In temporal terms, latency is the wait for the effects of the past to arrive in the present. As such, latency is a movement from past to present or even to future. It is the inverse temporal orientation of anticipation – in which the not-yet future reorients the present.[7] In comparison, latency in ecological time describes how the submerged sediments of the past arrive in the present to disrupt the reproduction of the same. Through latency, the future is already altered.

Latency and the St Clair River

Latency, I want to argue here, is useful when looking at the chemical infrastructures that are at stake for life in and along the St Clair River area. As the site of a chemical infrastructure, the St Clair River is not only a buzzing node in North America's petro-chemical network, it is also a landscape saturated with the effluent of industrialization. Since the 1950s, through layers of industrial processes, the 'excess' chemicals of production have been steadily moved into the waters, airs and ground of this region. At least since the 1950s, chemical dumping into the St Clair has been a source of public anxiety and scientific study. In the 1970s, in the wake of Love Canal, people also became increasingly concerned about the storage of industrial sludge and petroleum processing waste in salt caverns dug into the region's natural geology of underground salt deposits. Six hundred meters under the earth's surface, seventy-three caverns on the Canadian side of the river each hold the equivalent of three large surface petroleum tanks. Worry, too, has bubbled up over the great algae blooms that were strangling Lake Erie. In the 1960s, Lake Erie was declared a 'dead lake', as oxygen depletion caused by abundant algae had destroyed its ecosystem, killing local fish and plants. In 1969, an oil slick on the nearby Cayuga River, then considered the most polluted in the United States, caught fire in a location just before it pours into Lake Erie. Anxiety over the accumulated chemical alteration of waterways visible as sludge on the river's surface helped to prompt the first Great Lakes Water Quality Agreement of 1972, which was strengthened in 1978. The result, in the 1980s, was the pushing of effluent skyward, through stacks into the air. Today, the skies of Chemical Valley are thick with vapour, bright with flames of flares, and punctured with smokestacks decorated with twinkling lights.

Yet even in the 1980s, the past returned. In 1985, a toxic sludge congealment as big as a basketball court – known as the 'Sarnia Blob' – appeared in the St Clair River following a leakage of liquid industrial waste into the river from a cavern operated by Dow Chemical. The Blob was provoked by a leak of 2,500 gallons of perchloroethelyene into the river, which then congealed with chemicals from previous spills to form a sludge of arsenic, copper, cadmium, chromium, iron, lead, mercury, nickel, zinc, polychlorinated biphenyls (PCBs), hexachlorobenzene, phosphorus, manganese, oil, grease and a cocktail of at least thirty toxic compounds known as polyaromatic hydrocarbons.[8] Dow was fined $16,000 for the spill and then vacuumed up the sludge. Maxwell Cohen, chairman of the International Joint Commission that revised the Water Quality Agreement in 1978, described the latent presence of toxic chemicals in the area as 'the revenge of the industrial past and the chemical present'. That revenge is finding a variety of forms, he explained: 'Companies that were legally in business 50 years ago and dumped materials that now have proven to be lethal, or semi lethal ... No one controlled them 50 years ago, or 40 years ago, or even 30.'[9] The past defers its violence into the present.

In the contemporary time, river and lake sediment have become a contested site that materially archives this chemical past. To remove the sediment is to disturb it, and hence to release the toxic residues of the past back into animation. Yet the river and lake floors are not frozen in time – they continue to move and flow. PCBs, though banned in the 1980s, continue to move through the St Clair and Detroit Rivers and across Lake Erie, with intensive accumulations along the shores and inside canals. As a deep water shipping corridor, the riversides erode and the central channel is regularly dredged. Waste water continues to flow, and sediment is at stake. The chemical archive refuses to remain latent.

PCBs, along with mercury, have become the signature contaminants studied in these waterways. Both are persistent, and both were dumped in the river indiscriminately in the mid-twentieth century. Importantly, both can be studied because they are among the rare contaminants that are strictly regulated, with PCB production even banned. Their presence archives past regulatory actions, not the present, and thus do not call into question the effluent of the contemporary industrial frenzy. Maps of current PCB levels reflect the national origins of the studies that make them, the shores of Canada above or the US below become white space where data ends. Over time, PCBs flow down the river with the current into the Detroit River and into Lake Erie, a location marked with data points that shine red and purple to mark concentrations equal to or greater than a hundred times the 'Probable Effect Limit' of 277 mg of PCB per kg of sediment.[10] Other studies offer snapping turtle eggs as useful embryonic samplers of PCB levels in the waterways.[11] PCBs are reproductive toxins, giving their persistence a particular intergenerational form.

Informating Chemicals

Squeezed between Sunoco and Shell, the Aamjiwnaang First Nation is a small Ojibway group whose ancestors have long lived along the St Clair River. As a colonized people, the Aamjiwnaang First Nation (referred to in Canadian government documents by the bureaucratic name 'Sarnia Indian Reserve 45') lays sovereign claim to its small territory of 1,280.5 hectares wedged amidst the refining factories of major multinational corporations, such as Imperial Oil, Sunoco, Shell and the old polymer plant. In recent years, environmental justice activists at the Aamjiwnaang First Nation have worked with local doctors, scientists and lawyers to document the first known case of a dramatic reduction in birth ratio of boys to girls, not associated with a specific acute industrial or nuclear accident.[12] From 1999 to 2003, only thirty-five boys were born out of a hundred children.[13]

This site and community sits at the cross hairs of multiple scales of governmentality: that of First Nation governance, of Ontario and Canadian environmental regulation, of the US and Michigan governance that controls the other side of the river and of Ohio on Lake Erie, each involving different state agencies and histories. As the US's largest supplier of oil, Canada remains deeply invested in spatializing this region as a zone that is attractive to production and refinement. Moreover, it is the provincially owned coal power plant that contributes the most air pollution to Chemical Valley.[14] Directly on the other side of the river sits Michigan's twin coal-powered plant. Regulations do not prevent chemical emissions, but rather manage them in a setting fostered by the state as a designated zone of industrial activity. Regulation does not so much prevent emissions as produce a structure of quantitative practices for assessing limits, collecting data and making models. The anthropologist Kim Fortun calls this the *informating of environmentalism*, in which research and governance converge on questions of data collection, circulation and display.[15] The informating of environmental governance has become exemplary of neoliberal practices that seek to monitor chemical emissions rather than to enforce specific standards.

On the Canadian side of the St Clair, regulatory practices at the federal and provincial levels exemplify this informating of governance. Most regulation happens provincially, and the province of Ontario has generated lists of 'point of impingement' standards for some 300-odd industrially produced emissions. The 'point of impingement' refers to the moment an emission leaves the property line of an industrial facility. However, more important is the rule-by-permit structure of Ontario regulation, which takes its scale of regulatory concern as an individual industrial facility, to which it grants what is called 'Certificates of Approval'. Certificates of Approval are site specific, and hence weigh the economic costs of a facility and its existing equipment against the point of impingement standards. Through the Certificate of Approval, industrial facilities are individually given permission to pollute, and must generate annual reports about their equipment and success in reaching point of impingement standards, and file incident-specific reports about unapproved releases. In other words, through this permit process, pollution is permitted, managed and turned into data practices. This is what the province, on its Ministry of Environment website, calls 'doing business in Ontario'.[16] Importantly for the chemical infrastructure of the St Clair River area, Certificates of Approval are only facility specific and hence do not consider the possible cumulative effect to a region that has a concentration of industrial facilities.[17] In other words, regulations structurally make the intensity and multiplicity of sources of industrial chemicals in air and water invisible. The recently passed Open for Business Act promises to streamline this permitting process further.

At the national level, industrial facilities again have to file a report on their emissions of select federally listed 'toxic' chemicals, and the collective data is then published annually as the National Pollution Release Inventory, a largely informational exercise. The data in the informational reports are created by industry itself, or by for-profit environmental consulting firms that specialize in meeting this bureaucratic requirement. Such environmental consulting firms do a brisk business along the St Clair. In Chemical Valley, an industry consortium records ambient air quality levels with a set of eleven air-monitoring stations and makes this data available online. In contrast, the provincial Ministry of Environment runs only one such station in the city of Sarnia itself, some distance from the industrial corridor. Data production about this site, thus, is deeply contaminated with industry. The regulations form a discursive weave of data within which the materiality of industrial chemicals slip between, out the stacks, over fences and into imperception.

What is known about the chemical infrastructure of Chemical Valley also comes from Aamjiwnaang environmental activists, ecologists, local NGOs and transnational NGOs such as Global Community Monitor. Their many efforts attempt to document the complex and chronic chemical exposures that are almost entirely unmeasured by the Canadian government through such techniques as body burden analysis and grass-roots bucket air sampling. On a clear odourless day in March 2008, members of the Aamjiwnaang First Nation's bucket brigade gathered an air sample near their cemetery. Tested in the United States, carbon disulphide was present in the sample at 41 $\mu g/m^3$.[18] Neither the US or Canada has a federal air standard for carbon disulphide.

According to the Canadian Environmental Protection Act of 1999, carbon disulphide is not considered 'toxic'. This is not due to lack of evidence about the possible health effects of carbon disulphide, but rather because the 'mean' level of carbon disulphide that the population of Canada is exposed to as a whole does not exceed the standard of 'tolerable concentration', which has been pegged at 'the concentration for reduced peroneal motor nerve conduction velocity in a population of viscose rayon workers exposed' over a lifetime.[19] Thus, 'toxic' becomes a calculated regulatory condition, rather than a chemical property. According to the Canadian Environmental Protection Act, a chemical must first be present in a large enough quantity to constitute a danger.[20] The US, in juxtaposition, has an Occupational Safety and Health Association (OSHA) short-term exposure limit for workplaces of 36 $\mu g/m^3$ (or 12 ppm) in fifteen minutes. Diverse studies have pointed to carbon disulphide as a reproductive toxin, with exposures of 13 to 77 ppm shifting sperm morphology, decreasing hormone levels and causing menstrual irregularities in humans.[21] Another study reported increased spontaneous abortion at exposures of 2 ppm.[22] Carbon disulphide, which is carried away easily by wind, has an atmospheric half-life of twelve days, moves over property lines, and then winks out of existence, leaving in its place the lag between exposure and harm, wafting in a state of regulatory invisibility and air testing illegibility.

At the Aamjiwnanng First Nation, the violent effects of externalized chemical excess has been tracked in the last decade by locally crafted community health surveys, and body maps that have documented the embodied effects of living in Chemical Valley. It is difficult to sustain such efforts as it is hard to raise funding, and members of the Aamjiwnanng Nation must continue to find work in the plants of Chemical Valley. Researchers can arrive, collect and share data and then leave, not directly involving themselves in local environmental politics. While efforts like bucket sampling can catch snapshots of chemical exposures of the present, and community health surveys track health responses to the past, neither type of knowledge provides data with the necessary temporal arrow: that past exposure leads to later health problems. Environmental justice activists have learned the painful lesson that chemical injury is not just displaced *spatially* with super stacks, toxic trading and selective plant placement, but also that chemical injury is displaced *temporally*, such that accountabilities exceed the scope of individual lives, bioaccumulating or persisting over time, across regulatory regimes, beyond research grants and into the conjectural future.

Intergenerational Life

If the changed sex ratio of the Aamjiwnaang First Nation is the effect of past releases of chemicals, the temporal lag makes putting cause and effect together elusive. And this is the rub for chemicals that act as reproductive toxins or that can trigger cancers. The effects are not necessarily felt at the moment of the exposure, but later, in the future. Even if a chemical decomposes rapidly, winking out of existence, the effect it had on the organisms it encountered in its brief life might not be felt for years. Within the uneven spatial and temporal extensions of chemicals, people are 'living in prognosis', as anthropologist Lochlan Jain explains.[23] They are structurally living in the unknown individual probability of a statistical possibility, waiting for a symptom to arrive that can only point retrospectively and conjecturally to a possible harm.

In the case of endocrine-mimicking chemicals this latency is intensified, as endocrine-mimicking chemicals have a particularly intense effect on shaping developing fetal life – life not yet born, hence future life. Endocrine-mimicking chemicals, then, can manifest health effects in the next generation. Or the embodied response may not even be felt until the next generation, the grandchildren that will not be born. For example, research into the effects of the estrogenic chemical bisphenol A (BPA) on pregnant mice has found that the significant effects occur not so much for the fetus *in utero*, but for the eggs being formed inside that fetus, and hence effects are manifest for the potential grandchildren who will not be born.[24]

Here, the lives not born along the St Clair River may well be the effect of exposures endured by their grandmothers. Hence the sex-ratio effects experienced by the Aamjiwnanng First Nation may be the latent response of an acute exposure two generations ago. Or it could be the effect of continuous, multiple, accumulated multi-generational exposures crossing a threshold that has not become an epistemologically legible measure. Either way, the effects become a chemical manifestation of an ongoing colonial violence.

A 2006 national body burden study called 'Polluted Children, Toxic Nation' conducted by the NGO Environmental Defence included a family from Aamjiwnaang. Describing a world permeated by chemicals, in which collectively 'we are polluted', the study focused on sets of grandparents, parents and children, and tested the Plain family of Aamjiwnaang, composed of grandfather (who had the study's highest concentration of PCBs), father (who had the study's highest overall chemical burden), and granddaughter (who had one of the lowest overall chemical burdens, with only seventeen reproductive toxins detected). How to tally the intergenerational effects of this chemical infrastructure is a politically charged project towards making the layered temporal existence of industrial chemicals visible.

Amidst the cacophony of studies, a chemical infrastructure is only ever partially tallied, only selectively perceivable, despite the many ways data proliferate. Industrial chemicals cross between bodies, link generations, mix in the air, cycle through the water, blow out a chimney, seep through a crack, cross property lines, and haunt the sediment, yet all the regulations together do not make the chemical infrastructure particularly visible. The production of data as the primary product of regulation offers the fantasy that management is underway, that chemicals have been logged and as such contained. Corporate good neighbour reports by Imperial Oil, promotional efforts by the local Chemical Sustainability Alliance, and national toxic registries all work together to form the fantastical discursive effect of rendering Chemical Valley as unpolluted.

While work has gone into showing how the violence of this chemical infrastructure has concentrated in the space of Aamjiwnaang land (a parcel made available to such injury by a colonial history of displacement) it is also the case that the possibility of chemical injury exceeds the region of the St Clair River, traveling in commodities around the globe. The health effects of chemical infrastructures are also thus more ubiquitous in what S. Lochlann Jain calls 'commodity violence', in which the harmful effect of commodities are probabilistic for the population, but not causally isolatable or predictable at the individual level.[25] It is only as a commodity danger that BPA has become legislated against in Canada, not as an industrial ingredient.

In fact, the research that gave rise to the notion of 'endocrine disrupters' was sparked in part by the effects of plastic labware, particularly the test tube ingredient nonylphenol. Plastics, and the chemicals they leach, moved into laboratory practice, and in this way asserted their presence in cell biology research. Ana Soto and Carlos Sonneshein, whose early work on endocrine disrupters was prompted by plasticware in 1989, used this work to develop the first commercial endocrine disrupter assay.[26] In other words, the nonylphenol in the labware helped to materialize the test by which a substance is experimentally definable as an 'endocrine disrupter'. Soto and Sonneshein have extended their research into endocrine-mimicking chemicals as part of a more general theory about metazoan (animal) cell proliferation, and hence cancer. It has long been an assumption that metazoan cells are by default quiescent, while plant cells are by default proliferating. Hence animal cells, according to this understanding of quiescent, need to be stimulated to reproduce or become cancerous. Soto and Sonneschein, in contrast, argue that metazoan cells are also by default proliferant, but that there are cytoplasma inhibitors that halt their proliferation.

Endocrine, and endocrine-mimicking chemicals, can inhibit these inhibitors, and thus, by virtue of indirect negative feedback, release cells into a proliferating state. In this way, cancer does not become a problem of genetic mutations stimulating cells to proliferate but rather, of chemical agents interfering with inhibitors that prevent proliferation. In other words, proliferation is a latent property of animal cells.[27] Or put another way, latency is a crucial capacity that regulates cellular proliferation and reproduction. Chemical itineraries into plastics and then into bench science experiments, have prompted this lesson that changes the very epistemological orientation of cancer research and helps to give us the category of 'endocrine disrupters'.

Cancer is a serious concern in Sarnia, despite the fact that the city's relative prosperity stands in stark contrast to the decayed industrialism, shrinking population and poverty of a once-proud Detroit, past symbol of American capitalism. The good life of Sarnia has an unusually high rate of rare cancers. For men living in the community who have worked in the refineries, the overall cancer rate is about 34 per cent higher than the provincial average, the lung cancer rate 50 per cent higher, the mesothelioma rate (linked to asbestos) is five times higher and the asbestosis rate nine times higher. Recently, a Wilm's cancer cluster, a kind of childhood kidney cancer, has been identified on the United States side of the river. Wilm's cancer tumours have been correlated with maternal exposure to pesticides.[28] Cancers, with names and medical treatments, are just one way of living in prognosis within chemical infrastructures. Reproductive toxins are yet another, and unlike cancers, lives that are not born are largely uncounted and unnamed. There is no medical treatment for a past harm that happened before birth, and perhaps even before conception. There is no life to treat. The possible role of past exposures to endocrine-disrupting chemicals in the changed sex ratio among the Aamjiwnaang is difficult to prove. Nearby Sarnia shows no such change in sex ratio. Other known cases of human sex-ratio changes have been linked to acute occupational exposures or industrial accidents.[29] Almost no state funded research has been conducted to help study this question.

In contrast, the study of sex-ratio change and other effects tied to sex is flourishing in local freshwater biology research. On the St Clair, herring gulls have been found to have a reduction in males, and a greater portion of male embryo death. Snapping turtles are feminized. White perch have a 45 per cent rate of being intersex in the St Clair.[30] Non-humans can be dissected and even killed for the sake of developing better knowledge about the chemical infrastructure saturating the ecosystems. Their shorter life cycles make intergenerational harms easier to track and even accessible to study with laboratory-based experiments. Intersex organisms and changes in the ratio of males to females are bound together in studies that register such organism and population level alterations as signs of the unseeable chemical infrastructure swirling in the St Clair.

The round goby has emerged as a living sentinel for endocrine-disrupting chemicals in Great Lake waterways.[31] The round goby is a fish from Europe, first found in the St Clair River in 1990, likely transported there in the ballast water of ships. Since then, the round goby has rapidly come to occupy all five of the Great Lakes. Gobys are bottom feeders. In other words, they feed off the layer of micro-organisms, and particularly the mussels, that live in surface sediment. Mussels, in turn, are filter-feeding invertebrates known to bioconcentrate contaminants. Thus chemicals have found a new route of bioaccumulation in the round goby. At the same time, the round goby is so successful as an invasive species because it is relatively pollution tolerant. Nonetheless, round goby populations show signal sex changes in response to endocrine-disrupting chemicals. Ecotoxicologists capture goby samples in sites expected to have high levels of contamination and perform a variety of tests that cannot be done to human bodies, including dissection and measuring of the chemical load in tissue and livers, as well as the checking the condition of gonads. In particular, male urogenital papilla that are shorter in length, a skewed sex ratio (which has a rare higher rate of females in the Detroit River), and the presence of eggs and sperm in male goby testes are all signals of endocrine-disrupting chemicals.[32] So too is the presence of vitellogenin, an egg yolk precursor that is protein synthesized in the liver, which is not expressed in males unless they are exposed to xeno- estrogens.[33] In the last few years, students working at local universities have developed a PCR assay for the presence of vitellogenin in the round goby. Thus, the round goby, which lives only two to three years, is becoming a 'signature of contamination' organism for the Great Lakes, a living assay that provides a way to sample and map the presence of endocrine disrupters.

Fish play an important role in the chemical infrastructure of the St Clair. Commercial fishing was banned in 1970 because of the presence of mercury, one of the first environmental regulatory acts governing the river. Fishing was also a source of food for many who lived along the St Clair, including the Aamjiwnaang, and thus the Aamjiwnaang were among several First Nations communities who in the 1970s were diagnosed as having Ontario Minamata Disease, a form of mercury poisoning first diagnosed in Japan.[34] The Dow factory that sits directly upstream from Aamjiwnaang land has manufactured chlorine since 1947 using the mercury cell method, a process that released methylmercury (which is more toxic than mercury and readily crosses blood–brain and placental barriers) into the river. Elders remember playing with mercury in the local creek, and harvesting mercury to sell on the black market.[35] The loss of this traditional food source was yet another manifestation of colonial violence in chemical form. Aamjiwnaang residents are warned to eat only one fish per week, and pregnant women and children are warned to eat only one fish per month. Today, recreational fishermen along the river are surveyed about whether the fish they catch are 'tainted', and are asked to use their sense of smell and taste to evaluate fish safety. Dow is considering 'capping' the sediment along their stretch of the river with concrete in an effort to contain the mercury of the past.

Refusing to Remain in the Past

Chemical infrastructures have histories. But these histories refuse to remain in the past. Gaps in time, between chemical emission and effects, work to support the regulatory fantasy that the chemical infrastructure is safely contained as good business. The chemical industry calls this Reponsible Care™. However, on occasion the chemical infrastructure does spike into visibility with a tragic event – the actions of chemicals become acute and no longer incremental. Local industries and municipalities, as well as the Aamjiwnaang First Nation are all participants in the Chemical Valley Emergency Coordinating Organization (CVECO) put in place along Chemical Valley since 1952. The CVECO helps to prepare inhabitants for chemical disasters and coordinates emergency response among industries, running a radio channel specifically for announcing emergency codes. Since 1980, an annual Sarnia Area Disaster Simulation drill is hosted in Chemical Valley. Preparedness is exercised, and the promissory harm is glimpsed. A chain of emergency sirens run through Chemical Valley, including Aamjiwnaang land. When the sirens roar, residents are asked to 'Shelter, Shut and Listen', sealing themselves in their homes. Sirens are tested every Monday at 12:30, becoming a sonic reminder of anticipatory harms. Sirens too are part of the chemical infrastructure, auguring its possible future.

From the capacity to induce proliferation in cell biology labs, to the multi-generational reproductive effects that become a structural violence for communities, to the persistence of PCBs in turtle eggs, to sediments that won't stay still, to sirens warning of a future yet to come, chemical infrastructures are shaped not just by space but by time. A goby assay or a bucket sample can show recent or immediate exposure. A community health chart or sex-ratio count can show the bodily effects of an unknown past event. There remains a crucial temporal gap in knowledge-making. The multi-jurisdictional regulatory confluence at the St Clair does little to fill this gap. Instead, it secures a space 'open for business', in which data production can work to give permission to industries to continue to add chemical flows and drifts, with effects deferred into the future, and unaccounted for. The sediment, meanwhile, still holds many pasts releasable into the present. Time works its insidious loops in the toxic world, largely untracked. Chemical infrastructures are waiting for their many stories to be told.

6 MANAGING AN EVERLASTINGLY POLLUTED WORLD: FOOD POLICIES AND COMMUNITY HEALTH ACTIONS IN THE FRENCH WEST INDIES

Didier Torny[1]

> INRA thus feels that whilst soil pollution by HCH will no longer be an issue in 2010, it will nevertheless take several centuries for drainage waters to slowly cleanse the earth and put an end to chlordecone pollution.[2]

Introduction

What if policymakers took the fact that we live in a permanently polluted world[3] seriously, rather than considering pollution to be harmless, negligible or easily manageable? What would be the consequences of such an attitude? Since the end of the 1990s, the rediscovery of chlordecone in the surface waters of the French West Indies has given us a small laboratory with which to answer these questions by looking at how authorities have attempted to ensure that people live in safety. Chlordecone is a pesticide which until 1993 was used to fight banana weevils; it remains almost permanently in the soil and is only very slowly drained away. Whilst environmental pollution is not disputed as and when it is discovered, the issue of the health consequences of such pollution continues to be hotly debated. The toxicity of high doses of chlordecone is well established, due to the massive production of epidemiological and toxicological knowledge following the Hopewell factory disaster in 1975, leading to chlordecone being banned in the United States. The West Indian case raises the issue of its possible harmful effects (cancers, fertility problems) at low doses, following exposure over a very long period.

Of course, the question of low dosage has been raised for many different substances, from food additives[4] to radioactive elements,[5] and chemical substances have been a focal point of different standards introduced over the last fifty years,[6] in order to determine the conditions under which a substance is qualified as harmless. This regulatory science[7] is largely based on toxicological knowledge, shaped within laboratories, and in a few cases reinforced by epidemiological data. Conversely, knowledge from the field, focusing on actual exposure and its consequences, often remains controversial both in its methods and its results, even when produced by experts such as environmental health academics.[8] It is therefore often replaced by reductionist approaches[9] in the regulatory space, catastrophes and their subsequent trials being the main places where those different strains of knowledge confront one another.[10] Whilst social movements built around polluted sites or regions need science for their claims to be heard,[11] their aims are potentially more diverse, as the term 'environmental justice' suggests, and include notions such as the achievement of concrete results which directly impact at-risk communities.[12]

The chlordecone case involves all of the elements found in the literature, but within a very specific configuration based on two precise characteristics. First, it is the story of extensive pollution being gradually rediscovered through systems of measurement and not of a case which was triggered by the gathering of syndromes where popular epidemiology plays a significant role.[13]

Second, it is a case which is completely dominated by a single main actor, the French government, which through its different institutions and programmes plays every role from data gatherer and standard setter to norm enforcer and community health organizer. This paper analyses how France managed the pollution – accepted as lasting for centuries – and how local and global knowledge, from both the field and laboratories, are intertwined in order to limit the threat to human health while people continue to live in a contaminated environment. To this end, after describing the gradual discovery of the presence of chlordecone in different milieus and the associated health issues (water pollution, soil contamination, food contamination), we will focus on the question of standards relating to foodstuffs. We will base ourselves on three source types: first, various official sources, from Snegaroff's publication in 1977 to AFFSA (French agency for food safety) and its most recent opinions on nursing in 2008, along with reports from local consultation groups and parliamentary work. Second, sources from the actors who are either involved in the issue (associations, farming professionals, etc.) or who report on its progress (the local or national press). Third, we used elements reported by our contacts, especially with regard to an expert investigation in the French West Indies in October 2008.

From Invisibility to Full-Scale Pollution: Building the Tangibility of Chlordecone

The story of chlordecone is not one of the invisibilization of the pollution,[14] but of a re-visibilization process which began with water and which made the forgotten pesticide a tangible pollution.[15] This visibilization process also made it possible to retrace the history of the authorization and use of chlordecone between 1964 and 1993, a history which we will summarize before moving on to the rediscovery of its presence in the environment.

After four years of study in the French West Indies, in 1968 two companies requested authorization for sale in France under the Kepone brand name.[16] Within an international context of discussions on the persistence of DDT, the risks to the environment had already been highlighted by the Commission des Toxiques, which was in charge of assessing pesticide toxicity at that time, and the request was refused at the end of 1969. Nonetheless, two years later Kepone was granted a temporary authorization for sale whilst its competitors, fighting the banana weevil with HCH derivatives, had theirs rescinded. It consequently became the most used substance as from 1974. The Hopewell accident did nothing to modify France's regulatory attitude, and Kepone's authorization for sale was regularly renewed until 1980. It was only when the product became unavailable due to its production being prohibited by the Environmental Protection Agency that the French importer stopped requesting authorizations for sale.

After being severely affected by two cyclones in 1980 and 1981 and having exhausted their stocks of Kepone, banana producers had to cope with a proliferation of the banana weevil. Within such a tense economic context and without any reference ever being made to American data or decisions, in 1981 the Commission des Toxiques granted an authorization for Curlone, a new chlordecone formula. Upon request from French importers, Curlone was produced by a Brazilian company until the substance was banned in February 1990. However, the general provisions governing substances allowed sales to continue for a further two years; chlordecone was therefore used in the French West Indies until 1993.[17] It was at the moment this ceased that environmental monitoring of water was introduced.

Indeed, the presence of pesticides in the surface waters of the French West Indies had been investigated on various occasions as part of research projects, but it was as from 1991 that they were monitored on a routine basis in Martinique. Only some of the 500 substances approved in France for crop treatment[18] were scrutinized, which is to be expected given the fact that numerous crops found in mainland France are not grown in the West Indies. However, as the Balland, Mestre and Fagot report points out, 'some very commonly used molecules are not monitored'.[19] So whilst the results appear relatively satisfactory for the period in question, at least with regard to treated waters, the limited analytical capacities create two problems: it is impossible to continuously measure water pollution and it is impossible to measure the presence of certain molecules used on a local basis. In Guadeloupe, the situation is even more problematic, because pesticides 'are not monitored as part of health checks'.[20] Contamination is only measured when it is caused by accidents or when a study is carried out. This highly critical report thus asserts the need to limit sources of pollution, to protect catchments and to improve water monitoring, especially by developing local ability to do analyses, the aim of these steps being to ensure public health and the protection of the environment in a serious situation where the principle of precaution is explicitly mentioned. The conclusions of the report led to campaigns for stronger measures in 1998 and 1999, both in Guadeloupe and Martinique, each country having basin committees since 1996, said committees becoming the central arena for discussing the analyses. In this critical context, chlordecone was very much on the fringe compared to other pesticides.[21] It was only when analytical capacities were changed that it became a major subject of concern, standing out from most other pesticides whose contamination of surface waters was proven and worrying.

Discovery of the Chlordecone Problem: Waters Polluted by Previous Pesticides

By setting up a new partnership with a laboratory from the Drôme region in France, capable of measuring 220 molecules and of funding a major measurement campaign in seven catchments in 1999, the Martinique DSDS (Department of Health and Social Development) considerably modified the analytical landscape. At the same time, it also modified the understanding of the contamination of surface and underground waters, both raw and treated. The initial analyses showed high levels of contamination, particularly of HCH β (Hexachlorocyclohexane β), but the laboratory also discovered a 'peak' of contamination which required further analysis and which proved to be chlordecone, which is not searched for under standard practices and which is difficult to quantify. A similar study was carried out in Guadeloupe with the same results. In February 2000, shortly after the Martinique alerts had led to the closure of a catchment (July 1999), the results of the campaign to take samples from the Guadeloupe catchments caused a 'crisis'. It would seem that this was the first time that an impact on public health had been declared by a French authority. The result was a shift, from management confined to an essentially environmental issue, to the public management of a health matter. Several series of measures were implemented, starting with the water itself: closure of the most polluted catchments and restricted use of water in the most affected districts, management of network interconnections, distribution of bottled water, installation of activated carbon filtration units by those responsible for the polluted catchments and increased monitoring of water quality. These steps concerned not only drinking water from the tap, but also bottled water: the Capes Dolé company had to cease operation in order to install carbon filters and treat its spring water. After ensuring that no pesticides were present, it was allowed to restart production.[22]

A great deal of publicity (such as press conferences and televised debates) was given to these measures, as it was important to inform people about the water contamination and to change their habits where necessary, thus confirming that it was a legitimate problem. The fact that the risk was made public was largely due to the change of laboratory, which made the previously invisible problem tangible.[23] It was deemed to be even more worrying in that it revealed products which had been prohibited for a long time, thus raising the question of the duration of past contaminations and of previous works and practices. Whilst wondering about the use of Mirex, chlordecone and HCH, old studies which examined the state of the milieu, in particular Snégaroff's publication[24] and Kermarrec's report[25] were rediscovered, and in retrospect could be seen as early alarms or warning signs.[26] The continued environmental pollution being proven, the topic of discussion was thus water as a milieu and not just as a source of drinking water. The closure or treatment of drinking water catchments was the first and most visible action taken. This led to discussions on the question of drinking water standards and on the definition of 'pesticide residue': whilst some people equated 'drinking water' with 'zero chlordecone', the authorities replied that after treatment the residues in the distributed water were well below the standards in force. Here we have the classic argument on the issue of low doses that can be found in debates relating to other phytosanitary substances (atrazine and so forth) in mainland France – on the one hand a viewpoint with no toxicity threshold and on the other the defence of a model with a threshold, including safety factors. In other words, the question that certain associations were asking was: can one drink water containing very low doses of chlordecone throughout one's lifetime without any negative effects on one's health? The authorities' answer that these waters 'conformed to regulations'[27] did not satisfy anyone who felt that chlordecone is, by its very nature, dangerous or carcinogenic. The permanence of aquatic pollution in untreated water, followed by its extension to saltwater and freshwater fishery resources as from 2004, demonstrated the existence of a lastingly contaminated milieu and raised numerous questions as to the pollution spreading. During our time in Guadeloupe, we were thus able to see the concern of farmers from the Grande-Terre region, whose lands are irrigated by waters from Basse-Terre, with regard to the risk of seeing residual pollution gradually passed on to their own farms and produce. On the other hand, as we were reminded in relation to 'wild' springs, due to the invisible nature of pollution and to the 'natural' and non-commercial nature of water available in this manner, it is difficult to make people take notice of preventive publicity against its use. This can be seen in a report on Antilles Television in 2006, where women pay no attention to measures taken to prohibit the use of a spring: 'As you can see, the water is perfectly clear, so not polluted, and that's what we use to prepare our babies' bottles!'

Extension to Polluted Land: Spotting Contaminations

Research into the origin of the water contamination led to the issue of soil contamination and to questions of pollutants being transferred from soil to water. At first the question focused on the water: it was a case of measuring contamination levels and of understanding the mechanisms in order to protect the resource from an environmental and health point of view. The results showed contamination from organochlorine pesticides in the soil of banana plantations, even though 'neither the checks themselves, nor the examination of the different documents consulted on site, revealed any irregularities regarding the presence of any prohibited products'.[28] Whilst it exonerated the contemporary producers from any responsibility, the fact that the contamination had been there for some time did not alter the nature of the risk, and it was the land itself which quickly became the object of public attention, creating a second risk, relatively isolated from that of the water. This shift was properly seen for the first time in October 2000 during a meeting of Martinique's departmental Hygiene Committee, when the proposal for a stance on health, written by doctors and technicians from DDASS (Departmental Management of Health and Social Affairs) and the Cire unit (Inter-Regional Epidemiology Unit) of the InVS (Institute of Health Monitoring), recommended that:

> [I]t being likely that farmed soils previously treated with products with a long lifetime (HCH β and chlordecone) are contaminated, we must ask ourselves whether the animal and vegetable produce from said lands constitute a source of exposure to these molecules, and if so, to what extent.[29]

Thus, from being a source of water contamination, the soil becomes a potential source of food contamination. This was the reasoning behind the first samples taken for a study carried out in the first half of 2001 for Grephy (a regional phytosanitary group), in partnership with farming organizations: relying on the skills of the Drôme laboratory, they paired off foodstuffs with soils, for example sweet potatoes or dachines.[30] As from February 2002, the state of contamination of the soil, of the dachines, Caribbean cabbages and sweet potatoes was discussed in various institutions; it was also discussed more and more publicly, right up until the prefect's press conference in July. This gradual publicizing took place through a partnership between government departments and professional organizations and led to the creation of a legislative framework for the management of 'soil' risk: the 2003 prefectural orders.

These two orders, dated 30 March 2003 for Martinique and 30 October 2003 for Guadeloupe, were similar: before producing edible plants which grow in the soil, the soil had to be analysed. The details of the procedure, which is relatively complex due to the highly localized nature of the chlordecone pollution, is beyond the scope of this paper. Suffice it to say that the samples were analysed in specialized laboratories in mainland France. If they were positive, there was then an analysis of the plants produced, which, where necessary, were either destroyed (Martinique) or not sold (Guadeloupe). The objective was thus to restrict consumer exposure using a measure of an agricultural nature. Whilst the cost had to be borne by the producer, a system was rapidly introduced for the analyses to be funded by the European Union, thus making it easier for this health measure to be implemented.

At the same time as the orders, in Martinique the BRGM (Bureau of Geological and Mining Research) and CIRAD (Centre for International Cooperation in Agricultural Research) were tasked with developing a model for mapping the risk of soil pollution by organochlorine pesticides. A sampling plan for land plots was established on the basis of an analysis which used three main criteria: the history of the land ownership, parasitical pressure related to rainfall and the soil's capacity for retention. In particular, the analysis was based on three aerial photographs of the banana fields, taken in 1970, 1980 and 1992. The results showed a strong correlation between contamination and banana plantations on the one hand and the plantations' duration on the other.[31] Taking the model on the one hand and the results of the samples required by the order on the other, it was possible to define the zones of risk: 'There is no chlordecone in the soil in areas where chlordecone was not used: no detection of atmospheric contamination or of transfers via runoff waters. If the soil is sound, the plants are sound.'[32]

This result was very important because it made it possible to clearly separate polluted zones from sound zones, thus making it far easier to understand and manage soil risks. The possibility of the contamination being transmitted through repeated contact, anthropologically recurrent,[33] was not valid for the soils, even where they were washed by waters. On the other hand, the contaminated zones would remain contaminated for a very long time; no remediation was proposed, and people had to think in terms of 'centuries',[34] at least for certain zones. It would appear that the farming populations took risk measurement very much on board, because we have been told on several occasions that when land is now bought or sold, it is systematically analysed, in the same way that houses are examined for lead or asbestos. It must be pointed out that such analyses are performed outside of any regulatory framework – they mark the need to rebuild sound territories in a world where contamination is seen as being global. In the same way that Belarusian authorities tried to build in the decade after Chernobyl, it is a case of sharing information about what is dangerous and what is not, as the threat is invisible and intangible.[35]

The Food Issue: Governed by the Principle of Precaution

The land-related risk, which was environmental at this stage, was thus addressed on two fronts: first by constructing scientific mechanisms, whether geographical or agronomical, with which to define a risk probability grid for the contaminated zones/lands/produce. This information was public and was widely shared. The second front was individualized and related to information on a given plot of land or (in a far more limited fashion due to the cost and availability of analyses) on a batch of produce. This information was private and was not shared with the public. Nor was there any information system or any notices displayed onsite. It should be added that this distinction between sound and polluted was modified on several occasions, either due to additional contaminated lands being discovered, or because the definition of 'sound' was modified – we will look more closely at this in the next section. To the high or low probability of owning land previously used for banana plantations, must therefore be added on the one hand the level of the land's contamination (which can vary enormously over short distances) and variations in the soil-plant transfer, which depend on the soil and plant in question[36] and, on the other hand, the legal definition of sellable (therefore 'sound') land.

Whilst the issue of the transfer to plants was at the origin of the soil-related measures, it was not initially a very visible issue, despite public discussions in the summer of 2002 in relation to soil contamination, which we have just described. It was far away from the West Indies that the question of foodstuffs was revealed to a distant audience, three months later, in an article which appeared in the press in mainland France:

> one and a half tons of sweet potatoes, served with chlordecone, a highly toxic pesticide which has been strictly prohibited in France since 1990 ... But the prohibition must be strictly respected. As matters stand, there is no acceptable threshold or liberty to be taken with the law.[37]

The article stressed that the issue was governed by a system with no threshold, and that the simple presence of chlordecone required the produce to be destroyed.

In the French West Indies, the tests required by the orders relating to land led to measures of the same type: Martinique's plant protection department (SRPV) destroyed fourteen crops between 2003 and 2005 and the tests carried out on marketed goods by Martinique and Guadeloupe's DCCRF (Directorate for Competition Policy, Consumer Affairs and Fraud Control) led to the withdrawal of numerous batches due to non-conformity. The effect on producers was twofold: first of all there were major direct financial losses for those who were affected; second there were indirect effects on the produce's image. The environmental issue of water pollution thus gradually became more complex, with the addition of an agricultural issue and then a food problem. As more and more anti-chlordecone measures were introduced, the tension between health imperatives and economic logic increased. As far as drinking water was concerned, the existence of drinkability standards and of the technical mechanisms for achieving and checking them made it possible to reach a compromise that was representative of the regulatory system which had been put into place over the previous half-century. The situation was very different for foodstuffs, as there was no real data on residue standards.

In the absence of any data, contemporary changes to normative choices thus led French authorities to adopt an explicit principle of precaution which allowed no residue whatsoever, though of course this was dependent on the analyses being effectively carried out and on the extent to which chlordecone was detectable. To this initial stage of events must be added the search for a greater understanding in order to be able to establish standards based upon ordinary data used for expert opinions (Toxicity Reference Values – TRV, Maximum Residue Limit – MRL). The 'zero chlordecone' situation was considered to be a temporary one: in a durably polluted environment, it was now a case of standardizing the population's exposure to the foodstuffs. After the mobilization of the specialist mainland laboratories, it was once again actors far from the West Indies who were mobilized to develop the standards.

From Precaution to Standardization: Building MRLs for Chlordecone

Facing the peculiar situation of rediscovered pollution for which there was no applicable norm, the French government had to build new standards from scratch. Created in 1998 within the context caused by the BSE crisis and the arguments over GMOs, AFSSA was the institution tasked with developing these toxicological standards. It was asked to set an MRL higher than the detection limit of 6 March 2003, to assess the contamination of foodstuffs by organochlorine pesticides in Martinique and in Guadeloupe, and to come up with a TRV. The agency responded to these requests in an opinion published on 10 December 2003.[38] It had used existing scientific literature and the international classification of chlordecone (CIRC) to produce two toxicological references: an acceptable limit for repeated exposure and a maximum limit for acute exposure. But TRVs are theoretical references which do not allow one to directly assess the danger. This can only be achieved by measuring a population's exposure and assessing whether TRVs are exceeded, and at what frequency. In this section we will analyse, one after the other, two very different ways of doing the calculation, one using local data, the other using quasi-universal considerations, which produce two diverging regulatory systems.

A Standard Based on 'Local Data': Provisional MRLs for 2005

In the case of food, exposure depends on the level of contamination of the food and on the amount eaten. The levels of this type of exposure can thus be calculated by cross-referencing the food contamination data with those relating to eating habits (food buying, frequency of consumption and size of portions). To obtain the latter, an initial study (Escal) was carried out. This study[39] was a food consumption survey which was carried out in Martinique among approximately 2,000 people between December 2003 and May 2004, by Cire Antilles-Guyane, with help from other organizations (DSDS, Conseil Général, ORS and CGSS). In addition to the socio-demographic details procured, the survey gathered information on food buying and consumption. It was thus a very general survey which followed the approach of the first PNSS (National Health and Nutrition Plan) and which had the primary objective of improving knowledge at a time when there was an increase in chronic illnesses and conditions which were to a great extent linked to nutrition (obesity, diabetes and arterial hypertension) and which are very prevalent in the West Indies. The objective of measuring exposure to chlordecone was integrated in an opportunistic fashion into the survey, with particular focus on population groups who ate significant amounts of local produce and who might therefore be contaminated.

At the same time, the monitoring and testing plans introduced by the various administrations involved, as mentioned above, made it possible to analyse root and non-root vegetables, meat and fish, fruit, seafood, milk and eggs.[40] Despite the resources required (refrigerated transport to mainland France), a large number of analyses were carried out, but with the exception of root vegetables the number was nevertheless too low to be deemed representative. There were thus demands for additional specific surveys, leading to the Reso survey starting in September 2005. However, the uncertainties and limits of the contamination surveys did not prevent AFSSA from publishing its report in a context of 'urgent requirement for the populations, local actors and public authorities'.[41]

By cross-referencing the two series of data described above, AFSSA made a theoretical assessment of the extent of the exposure by determining a theoretical maximum daily amount, based on local produce. This approach was based on the hypothesis that all of the foodstuffs of local origin which had been consumed and which might have been contaminated were contaminated to a maximum extent. According to this maximalist method, it appeared to be possible to cover the risk of going over the TRVs for all age categories. However, whilst this scenario indicated the maximum limits which could be tested, it overestimated exposure and was unable to identify the main foods which carried chlordecone and which should be regulated as a priority. Other methods were then used and were able to indicate that the food group carrying the most chlordecone was root vegetables. Results showed that approximately 7 per cent of the individuals included in the Escal survey (108 people out of 1,495) had higher TRV levels than those recommended by AFSSA, thus justifying the setting of limits for foodstuffs. The report also pointed out the problem of own consumption, given that only marketed foodstuffs were going to be tested:

> In order for the exposure of own consumers to be effectively reduced to acceptable levels, it would appear necessary to combine the setting of MRL with consumption recommendations for own consumption, especially with regard to produce from contaminated zones.[42]

This was the basis on which the new produce standards were defined, in the form of two orders which introduced the agencies' recommendations:[43] foodstuffs of animal origin (5 October 2005) and foodstuffs of plant origin (10 October 2005). The logic behind the standards distinguished between two types of produce, in accordance with their consumption and their contamination: those which were the biggest vectors must have a level of chlordecone below 50 µg/kg, the remainder must be below 200 µg/kg. This creation of MRLs was presented with caution, due to the uncertainties surrounding contamination (greater than those surrounding consumption). Whilst overall they were well received, after a long wait, especially by certain farmers in contaminated zones who would once again be able to produce certain crops, these initial MRLs were criticized, both for their uncertain basis – backed by the principle of precaution – and for their very theoretical principle. Using the standard criticism of the cumulative effect of substances, this new standard, specific to chlordecone, was a decoy.[44]

Two new surveys provided further knowledge: on the consumption side, Calbas, which replicated Martinique's Escal survey in Guadeloupe; on the contamination side, the Reso survey, developing analyses for numerous foodstuffs. Carried out between November 2005 and July 2006, Reso broadened the range of crops taken into consideration (cucurbitaceous plants for example) and made it possible to analyse plants which were not (or not to any great extent) monitored by government departments (especially own consumption and short circuits). As with the other surveys, Reso was widely presented and discussed in consultation groups such as Grephy and Grepp, and its results were expected. On the basis of Calbas and Reso, AFSSA re-examined the issue of the 2003 TRVs and the 2005 MRLs.

Overall, the estimated levels of chronic exposure among the French West Indian population were lower than those previously assessed, due to the contamination data used being more representative. The Reso surveys were carried out using a random sample of foodstuffs available within distribution circuits, whereas the monitoring and testing plans which had initially been used more particularly targeted foodstuffs and zones likely to be a risk. In addition, AFSSA considered that these results integrated 'the effectiveness of risk control measures already implemented: the prefectural orders issued in 2003 with regard to the preventive analysis of soils, the maximum limit of 0.1 μg/l in drinking water, the provisional maximum limits for food contamination'.[45] However, largely due to own consumption, the agency believed that up to 18.5 per cent of children aged between three and five years, and living within contaminated zones, might have gone over the acceptable limit for chronic exposure. So it did not change the MRL, but suggested adding marine produce and Caribbean cabbage to the list of foods with a limit of 50 μg/kg. After a lengthy and costly process of data collection and major management steps, the situation thus appeared to have stabilized, both with regard to the risk itself and its standardization. Yet AFSSA's proposals were not implemented, because at the same time another legislative process was underway.

A Standard with Universal Scope: The 2007 European MRLs

In 2006 the European Commission began a process of harmonizing the standards for pesticide residues in force in the different member states of the European Union, in accordance with regulation 396/2005/CE of 23 February 2005 relating to the 'maximum limits applicable to pesticide residues present in or on foodstuffs and animal feeds of animal and plant origins'. As chlordecone is listed in annex 3 of this regulation, the French authorities had to suggest an MRL to the Commission. Initially they tried to escape this legislative framework by declaring that chlordecone was not (or was no longer) a pesticide. In a letter to DG Sanco (Directorate General for Health and Consumer Affairs) dated 5 December 2006, they asserted that chlordecone was an environmental contaminant – just like the mercury used by the wildcat miners in Guyana or the heavy metals produced by factories – and that as such it should be removed from the regulation.

This approach was hardly crowned with success because the European Food Safety Agency (EFSA) based itself on the MRLs recommended by AFSSA to closely examine the case of chlordecone. Whilst AFSSA had based its limits on the specific study of contaminations and consumption in the French West Indies, EFSA projected MRLs for the twenty-six European food regimes, which allowed it to model population exposures, and its conclusions were irrefutable. Whilst the foodstuffs were all contaminated to MRL levels, the TRVs had been significantly exceeded for the entire population of Europe. As a result, the MRLs could not be accepted. The differences between the two agencies thus stemmed from the ways in which exposure was approached, with on the one hand a realistic model based on the data collected, and on the other hand a theoretical model.[46] AFSSA and the French authorities argued that the pesticide had not been used for fourteen years, and that therefore it was unrealistic to imagine any generalized contamination at MRL level; the European authorities gave France a few months to come up with a full report on the chlordecone MRL.

In the summer of 2007, the chlordecone case changed in nature: the complaints registered in the French West Indies for 'poisoning' began to be relayed at national level and the case was compared to the tainted blood affair. Upon request from a local association, cancerologist Professor Belpomme compiled a report on the consequences of chlordecone pollution and announced that it would be presented to the French national assembly on 18 September. The media went crazy, first of all when the Ministry of Agriculture declared the situation in the West Indies to be 'very serious' and then when Professor Belpomme amplified matters by talking about a 'health disaster' and a situation which was 'worse than that of the tainted blood affair'. Even though the Minister of Health announced the following day that there was no 'scientific proof', she did call upon people in Martinique and Guadeloupe 'to take the greatest care with regard to produce which does not come from commercial circuits (produce from family vegetable gardens or water taken directly from springs)'[47] and above all announced a new MRL value of 20 μg/kg, thus demonstrating the government's determination to strengthen standards by reducing the threshold. Unlike other cases where it had denied that there was a problem or that there were potential consequences,[48] the French government increased concerns.

However, this proposal for a universal MRL did not convince EFSA and the commission, who considered that produce cultivated outside the French West Indies and which had no reason to be contaminated should be standardized by the limit of analytical quantification. On 24 October 2007, the European Commission accepted MRL values lower than those defined in 2005: 20 μg/kg for crops cultivated in tropical or temperate climates (citrus fruit, tropical fruit, all vegetables, corn, sugar cane and so forth) and 10 μg/kg for certain plants specific to regions with temperate climates or which could be imported from countries other than the West Indies (wheat, rice, apples, pears, fruits with stones, sugar beet and so forth). Whether they came from land or sea, animal foodstuffs had to respect a limit of 20 μg/kg. These measures were set out in the order of 30 June 2008, which replaced the 2005 orders. Between 2004 and 2008, there had thus been three different MRLs for the same produce: at first root vegetables could not contain any chlordecone, then they could contain up to 50 μg/kg, and finally they could contain up to 20 μg/kg. Far from 're-establishing consumer confidence in local produce', these changes in standards had a negative effect on the credibility of assessment measures and risk management, and came in for considerable criticism. Some farmers felt them to be too strict, seeing them as 'excessive precaution' or as an 'umbrella policy'.[49] Symmetrically, the dual standard of 20 μg/kg for local produce, as set against 10 μg/kg for produce from mainland France, fed the theory of governmental laissez-faire or of the authorized poisoning of West Indians, in a period which was marked by health scandals. Some people made repeated demands for MRLs to be removed, which would mean dropping all foodstuffs to a limit of 10 μg/kg, which is the currently accepted quantification limit, and which would take matters back to the situation prior to 2005.

So over a period of three years, some products were forbidden, then authorized, then forbidden again as the normative frameworks based on different assemblages of knowledge shifted. From the field perspective, not only is it very difficult to understand, but it also undermines belief in the accuracy of such norms, these changes being interpreted as the result of political struggles. If such struggles did exist, we should also highlight the points they had in common, such as the quantification limit of 10 μg/kg, which is applied to 'standard' but not to 'advanced' equipment. If the European Commission had taken its reasoning to its logical conclusion, it would not have imposed a precise limit, but rather one which depended on the equipment available. Furthermore, this type of limit is only valid for commercialized products, as these sanitary norms are deeply entrenched within a made-for-market system.[50]

Taking Care of the Household Products: The JAFA programme

The story of chlordecone could end here, being just another episode of the standardization of pollution through regulatory science imposed by experts, governments and economic actors. Nevertheless, the tasks usually undertaken by NGOs on behalf of the population[51] were in this case carried out by the government, which went far beyond traditional risk communication. How did this happen? In order to understand, we have to go back to the food inquiries which were instrumental in building the first MRLs: they also showed in 2006 that the populations at risk were mainly those who consumed home-grown vegetables cultivated in contaminated soils. Getting in touch with this population, in a context of very heterogeneous pollution and land use, was no easy task and it was indeed costly, and thus highly improbable in a French context with little tradition of community health.[52]

Reusing Provisional MRLs Knowledge

The context of the health scandal from the autumn of 2007 onwards was the triggering factor that made this improbable programme possible. In the 2008 National Chlordecone plan, EUR 3.2 million was allocated to build a community public health programme 'without parallel in France'.[53] This programme was named JAFA for *jardins familiaux* (family gardens), as the traditional Creole household garden was seen as the main source of risk. It aimed at identifying the households potentially overexposed to chlordecone and helping them reduce their exposure by different means.[54]

Initially the programme had to rely on distant knowledge of the contamination and reused an updated version of the geographical modelling of the risk based on banana cultivation developed in 2003, including the level of precipitation – correlated with the frequency of pesticide use and the nature of the soil. Limiting the zones to be examined, JAFA promoters crossed the probabilistic model with land ownership maps to come to a figure of 17,887 lots. The fieldwork began with public meetings in all towns concerned, providing information on chlordecone, environmental pollution (water, soil) and its potential sanitary impacts through food, followed by the distribution of leaflets through mail boxes to inform people of the launch of the door-to-door campaign. Fifteen people were hired and given a week-long training course on chlordecone, the JAFA programme and handling face-to-face inquiries. They then had to go from lot to lot to find the families who owned and/or cultivated them, suggesting later appointments if they were not present. Despite this flexibility, 36 per cent of households in risk zones were not actually interviewed.

Where they succeeded in contacting the families, they handed out a questionnaire asking about the existence of or desire for a household garden, the source of their vegetables and the frequency of their consumption. Once again, the knowledge used for MRLs was reinvested: the threshold was set at a twice-weekly consumption of root vegetables, coming from their garden or that of their neighbors. When they were above this threshold, as was the case for 20 per cent of the households, in order to determine the exact levels of chlordecone, a soil analysis was proposed and then processed by local and mainland laboratories within two months. Then, taking a theoretical 20 per cent rate of transfer from soil to plant calculated for farmers' products in 2006, the JAFA field workers recommended actions depending on the level of contamination. It went from no restriction of consumption for low levels of chlordecone, to a reminder that it was forbidden to sell the vegetables, and on to a complete stop. The final result was that less than 600 people were affected by such limitations and, as such, really considered to be 'at risk' by official standards.

The approach of the JAFA programnme is entirely based on an MRL type of knowledge, but built on the 'local science' of soils, plants and consumption. It is an extension of such standardized knowledge to very local conditions – one household, one lot. Contrary to the ETHOS programme developed for post-Chernobyl contaminated territories, it is not a participatory programme:[55] the aim is less to promote local initiatives and to give people new means to cope and live in their environment, than to treat them with a slightly adapted global standard. As professional farmers, they now know the status of their lot, the only difference being that in some cases, they are 'authorized' to eat food that is not marketable. But in the top-to-bottom JAFA model, new knowledge can only come from academic expertise, whether on soil remediation or plant transfer.

The Limits of Standardized Management

The extension of quasi-MRLs to household products needed a lot of material and human investment, at an unprecedented level for the French public health system. The model nevertheless showed its limits by ignoring three important situations. First, there was a limit by design: as the promoters of JAFA themselves acknowledge, some households outside the defined 'risk zones' could be under the threat of chlordecone, as it was used not only for controlling banana weevils, but also for other agricultural uses, including low-scale farming. This population must be added to the 'non respondent' households. In a very classic way, promoters consider the hypothesis leading to the recommendations to be very pessimistic, thus stating that 'the population actually over the TRV is probably inferior to the one estimated in the JAFA programme.'[56]

The second limit is linked to the behavior of chlordecone. As far as the soil is concerned, it is an incredibly stable molecule: contamination is very localized, thus allowing safe use of the vast majority of the land. If we return to the issue of water transporting chlordecone down the rivers, we find another area of contamination: fish and shellfish. Indeed, the situation is a worrying one, as fishing was forbidden in certain rivers from 2004, shellfish farms were shut down by authorities from 2007 onwards and high levels of contamination were found in some animals.[57] Soil leaching models have shown progressive contamination of coastal seawaters. So with their high levels of self-sufficiency, fishing villages were identified as being at risk. And unlike vegetables from Creole gardens which proved to be less contaminated than expected, seawater is a milieu which homogenizes pollution, thus potentially condemning populations to radically change their consumption habits, something which from an economic standpoint could be costly. Moreover, it may be bad for their health as replacing current intake with imported products might increase the rate of metabolic diseases even further – especially diabetes.[58]

The third limit is a socio-political one: vegetables have only been considered in two possible contexts, either as a commodity submitted to MRLs or as a home-grown product 'controlled' by the JAFA programme. In fact, there is a very pervasive third context: products sold illegally as part of an informal economy, usually called 'roadside markets'. As it is illegal, no-one has any real idea of its size, though small 'shops' can be seen everywhere along the roads in the French West Indies. Anecdotal evidence also exists:

> We had found a cow with lots of chlordecone, so we went to the farm, which was very small. The old farmer was there with his three cows, next to a stream which was clearly the source of the contamination. And suddenly, on the other side of the stream I saw a huge field of ignames and I asked who the owner was. Nobody knew. I checked the land ownership map, and it was national public domain! An illegal crop of such a size was certainly for the roadside market.[59]

This officer had a hard time obtaining authorization for the crop to be analysed (and maybe destroyed), as 'it was owned by no one, so not destined to market'. The informal economy is crucial to some part of the population and authorities know it, so crucial that even such a mobilization for health has not changed that. Contrary to other policies such as drug enforcement or advertisements for condoms that have been transformed by the AIDS epidemic, it has not been yet sanitized[60] and remains a larger socio-political issue.

Conclusion: Living in a Polluted World

For two or three decades, French West Indians were totally unaware that their environment was being contaminated by chlordecone. Drinking from a spring or from a tap, eating vegetables or buying them from a local market required no particular vigilance. The gradual revelation of the state of pollution demonstrated the poison's invasive nature. Just like that of the prion (meat products, but also blood products and medical devices) or radioactivity (mushrooms, thyme, marine produce and so forth),[61] the gradual circulation of pollution throughout the production, processing and distribution networks created concern and justified legislative action. The realization that land would be gorged with poison for centuries, and its steady progression via leaching of the soil to marine produce, was coupled with the realization that the human body could be saturated with the poison too, with pregnant women and young children being especially at risk. The effect on pregnant women and their offspring was studied in cohorts by epidemiologists.

Whilst we already know that MRLs are as much a political tool as a scientific apparatus for treating pollution,[62] the chlordecone case shows this through the confrontation of two legitimate authorities and their MRLs settings: the French government and the European Commission. But it also begs the question for whom MRLs are made. The pre-2005 precaution caused the ruin of numerous small farmers and threatened consumption of local produce in favour of food regimes which were more expensive and sometimes less nutritious. EFSA found an answer for European consumers, but did nothing to solve the problem for the West Indians, even giving the latter the impression that mainland Europeans were being protected to the detriment of the health of West Indians. The AFSSA standards appeared to be a compromise, but left aside the health of young children and families who were generally at a socio-economic disadvantage and who relied to a considerable extent on own production (vegetable gardens, fishing). It was not directly through social movements, but rather through the threat of a large-scale health scandal that a community health programme was developed for them, extending a quasi-MRL regime, but leaving aside other populations (fishermen and those using the informal economy). Nevertheless, a new three-year plan (2011–13) takes the seawater pollution and the fishing villages into account, and is designed to produce knowledge and ensure surveillance.

Though it has taken public authorities by surprise, the rediscovery of chlordecone has not led to any regulatory crisis.[63] Even though it took time and different options were explored, it is the regime defined by MRLs which was used to manage the problem of continuous low-dose pollution. Nevertheless, this regime is potentially threatened by two factors. On the one hand, all of the measures were introduced when confronting the potential health effects of the pollution, and not any proven ones. This is not the case anymore, as chlordecone exposure has been epidemiologically linked to prostate cancer,[64] which was already very high in Caribbean populations. On the other hand, as soil remediation is not yet considered possible and the pollution will last for centuries, it could lead to constant questioning of pollution which exists and is recognized in a very pure form, as the French West Indies have almost no major industrial pollution. A laboratory island, with one substance and human populations living together with no means of escape.

7 CHERNOBYL EMPOWERMENT? EXPORTING 'PARTICIPATORY GOVERNANCE' TO CONTAMINATED TERRITORIES

Sezin Topçu

Introduction

Citizen participation in the management of science and technology became a key component of public and collective action. The participatory ideal is now mobilized at almost every level, by NGOs, watchdog groups, trade unions, by techno-entrepreneurs, science policy proponents, EU bureaucrats and even by candidates to presidential elections.

Since the end of the 1980s, the STS field has played a pioneer role in the promotion of public participation as a modernized way of governing science and technology.[1] Most of the literature produced within this context was based on the hypothesis that more public participation would necessarily lead to the empowerment of public and to a democratization of science. This paper questions such correlation. Empowerment is analysed, in this frame, as a vague, even 'powerless' notion (despite the idea of power it incorporates), given the fact that it was largely 'normalized' by international bodies and industrial actors, thus differing from its first theorizations by the social movements (social workers, feminists, development NGOs, trade union movements ...) as a means to transform the social order and power relations.[2] In line with recent critical work calling for a deconstruction of institutional discourses,[3] for a critical analysis of the official framings,[4] and for a re-evaluation of the participatory 'ideal' in relation to the politico-economic context that gave birth to it,[5] this paper takes seriously the cases in which public participation does not necessarily allow a democratization of science, but rather serves to increase the social acceptability of a controversial innovation,[6] to restore public trust vis-à-vis science,[7] to dictate narrow frames of 'debate' on technological innovation[8] or to shift towards a lesser commitment of the state in public health protection, thanks to an increased commitment by 'empowered citizens'.[9] In particular the latter dynamic is emblematic of neo-liberal governmentality as analysed by Foucault[10] and Foucauldian theorists. Neo-liberal 'art of government' refers to a specific set of strategies, instruments and discourses through which individuals are governed and are expected to govern themselves, thanks to a relative freedom of action attributed to them. It aims, among other things, to transfer responsibility to individuals about the social problems/risks with which they are confronted.[11] Competitiveness, entrepreneurship, autonomy, self-achievement and self-responsibility appear to be the *mots d'ordre* promoted within such a setting characterized by market rationality and economic calculation.[12] Whereas the state pretends to 'govern as little as possible', public participation and empowerment are put into the agenda as novel technologies of self-government.[13]

To analyse the ambivalent character of participatory practices, this paper focuses on one of the oldest controversial innovations of the contemporary period, namely the nuclear power technologies which provide a rich historical insight on these questions. In France, the most nuclearized country in the world with its fifty-eight power stations, institutionally organized public participation is still a marginal form of public action. The technocratic decision-making which characterized the nuclear sector[14] has not really disappeared although the nuclear authorities publicly affirm their attachment to dialogue with the public, even pretending that a 'cultural revolution' is currently at stake in nuclear–society relations. In practice, public participation does not find homogeneous acceptance among policymakers, experts from regulatory bodies or industrial entrepreneurs.[15]

In France and in Europe, post-accident management is one of the basic fields in which public participation in nuclear energy has been experimented and promoted. The management of territories contaminated by the Chernobyl accident has been a big concern for international bodies, nuclear industry and decision-makers for the last twenty years. Western experts were regularly mobilized for field research and observation to judge health impacts and to formulate recommendations for the Ukrainian, Russian and Belarusian governments. For the last decade, 'participatory governance' has become the new trend in this field. Innovated by a group of French experts under the banner of the 'Ethos approach', it was implemented in Belarus, the country most affected by the Chernobyl disaster, with 25 per cent of its territories classified as highly contaminated. Participatory attempts rapidly gained the support of the European Commission, the Belarusian authorities and international organizations such as the World Bank and the OECD. Gradually, the Ethos project became a reference for establishing post-accident management systems for Europe.

What does public participation mean in a context of long-term contamination? In what ways does it transform traditional risk management? How is the balance of responsibilities between state and public redefined within this framework? What does autonomy of individuals *vis-à-vis* state institutions and experts signify in cases of strong economic insecurity? To what extent does empowerment of the public in contaminated territories equate to a democratization of nuclear–society relations? What other roles does it play?

To tackle these questions, I first present the background of the Ethos project, by describing the nature of the initial rehabilitation projects implemented in Belarus and by showing how state commitment for the protection, relocation and compensation of victims was gradually challenged by the technical complexity and the high economic cost of radioactive risk management. I then analyse the basic assumptions, tools and discourses associated with the participatory projects conducted in Belarus starting from 1996. I explain how and why these projects turned out to be a success story at a European level. Finally, I treat the reactions and the social–moral dilemmas generated by these projects in the NGO arena. I argue that public participation in post-accident management aims, above all, to transform 'victims' into 'productive' actors urged to shoulder, in part, the state's burden in managing the risk society. I propose to call this process the 'politics of empowerment'.

A Growing Disaster Uncovered

The demise of the USSR led to a relative opening up of information on the consequences of the Chernobyl accident as well as an internationalization of efforts for the management of the contaminated territories. One of the major political questions then concerned the compensation and the relocation of the inhabitants of these territories.

Until 1991, the category of contaminated zones, outside the immediate evacuated zones, was determined by reference to surface contamination by Caesium-137. In these zones, a threshold life-time dose, called the safe living concept, was determined for the public's exposure to radioactivity, the object being that doses above this value should lead to evacuations. In 1990, the threshold dose was 350 mSv (35 rem), a value five times higher than the maximum authorized dose defined by the International Commission for Radiological Protection (ICRP) during the same period. Despite the fact that it had been rendered much more restrictive since 1986, the Soviet radioprotection system of 1990 thus excluded many victims, most of whom were living in zones considered contaminated according to the international norms.

During the first days after the disaster, around 115,000 people living within a perimeter of 30 km around the Chernobyl reactor were evacuated. This region, now called the 'dead zone' (*zone morte*), including two cities – Pripyat and Chernobyl – and seventy villages, was permanently evacuated. A few months later, evacuations were extended to a perimeter of 50 km. Without any doubt, prior to 26 April 1986, none of the inhabitants of these regions had ever imagined that their home would one day become impure, that all the land around their villages would be burned, and that even access to the family cemetery would be forbidden forever.

Indeed, during the initial phase, outside the immediate evacuated zone, the official doctrine was a rapid 'return to normal'. The government intended to liquidate the consequences of the accident by decontaminating the soil, by opening it up to agricultural activity, and by repatriating the evacuated populations. Even the experts did not imagine that areas remote from the reactor could have become highly contaminated.[16]

Nevertheless, the optimism of the experts and the official secrecy, strictly enforced despite the glasnost,[17] were soon set aside by the immense harm caused by the disaster, which made it difficult to conceal. Massive efforts to get rid of the contamination by the removal of the soil, the daily hosing of the roads, asphalting, demolition of contaminated buildings or burial of the wastes did not resolve the problem. Plans to drain the contaminated lands turned out to be ineffectual. Preventive acts even proved to be counterproductive as they ruined the fertility of the soil. They also appeared to be pointless since the treated areas were re-contaminated by the rapid transfer of radionuclides from neighbouring land. Thus, gradually, the permanent character of the radioactive contamination became undeniable in the most affected zones. In Belarus, between 1986 and 1989, the state banned agricultural activities in 257,000 hectares. Categorized as contaminated were 1.6 million hectares of agricultural land and 1 million hectares of forest.[18]

From 1989, thanks to the removal of the official censorship on the consequences of the Chernobyl explosion, a real media debate developed and helped to transform the accident into the 'most terrible disaster of the 20th century'.[19] The increase in illnesses and cancers as well as the observation of anomalies in newly born animals were massively reported. Previous accidents, which had been concealed until then, in particular the thirty-two-year-old Kyshtym accident (1957), were made public. The 'Russian civil society' became more visible within this process.[20] It demanded a revision of the rehabilitation strategies, the complete shutdown of the Chernobyl reactor and the abandonment of the power stations under construction. Scientific controversies also became public. In April 1989, the Belarus Academy of Science rejected the Soviet 'concept' of 350 mSv, advocating the adoption of a safer norm, i.e. 70 mSv, which corresponds to 1 mSv per year.[21]

In particular, the discovery of new – and very significant – contamination in areas situated well beyond the 30 km zone, in south-east Belarus and in south-west Russia, catalyzed the controversies and the protests. Not only did the evacuated local populations have to accept the idea that they would never return to their homelands, but also the prospect of new evacuations estimated to involve another 200,000 people started to gain weight. Those believing they had survived the disaster thanks to their geographical advantage suddenly became aware of their vulnerability. The biggest shock was felt in Belarus, a non-nuclearized country but nevertheless the one most touched by the Chernobyl accident. Gradually, it would be revealed that 70 per cent of the land contaminated by the explosion belonged to this country. A quarter of the territory of Belarus proved to be highly contaminated, a situation which led the Belarusian government to declare its national territory a 'zone of ecological disaster' in July 1990.

In October 1989, Belarus adopted a new evacuation plan for many villages, some of them up to 250 km away from the Chernobyl reactor. The same year, the government of Ukraine, where almost 9 per cent of the national territory was categorized as contaminated, decided to evacuate fourteen more villages. Certain villages which had been used, at the beginning, as transfer and refuge areas for the inhabitants of the initially evacuated zones, turned out, a few years later, to be too contaminated to be habitable, such as the Ukrainian village Roudnia-Ossochnia.[22]

For all these reasons, the loss of public confidence in the authorities reached a peak in spring 1989. During the big public demonstration organized in Moscow on the third anniversary of the disaster, demands for the establishment of an independent expertise under the aegis of international organs were massive. The International Chernobyl Project was launched by the International Atomic Energy Agency (IAEA) within this context, through an initiative of the USSR government, with the intention of studying the radiological impact of the accident and of evaluating the appropriateness of the protective measures adopted by the Soviet authorities.

However, the conclusions of the project, published two years later, turned out to be highly disappointing for those who were expecting an improvement in their living conditions and their health, through the intervention of international bodies, then considered as neutral mediators. Conducted by 200 experts from 25 countries, with a dozen French experts (some of whom innovated the participatory Ethos project several years later), the International Chernobyl Project judged, in 1991, that the sanitary impact of the accident on the surveyed populations (i.e. several thousand people randomly selected in thirteen villages) was negligible.[23] Indeed, the damage engendered by the disaster was branded as mere psychological problems by the experts mandated by the IAEA. The only exception to this was the brief mention, in the final report, of the possibility of a slight increase in the number of tumours of leukaemia and of the thyroid. Just like the World Health Organization (WHO) who had considered, two years before, that the threshold levels fixed by the Soviet authorities to determine evacuations were 'too cautious',[24] the IAEA judged that the measures taken by the USSR after the accident (in particular the food restrictions and the evacuation plans) were 'excessive', as indicated in the final report :

> The protective measures taken or planned for the longer term, albeit well intentioned, generally exceed what would have been strictly necessary from a radiological protection viewpoint. The relocation and foodstuff restrictions should have been

> less extensive. These measures are not justified on radiological protection grounds; however, any relaxation of the current policy would almost certainly be counter-productive in view of the present high levels of stress and anxiety amongst inhabitants of the contaminated areas of concern and people's present expectations.[25]

Relocation: A State Policy Transformed into a Utopia

The quite optimistic report from the IAEA did not much influence the decision of the newly independent ex-Soviet republics to adopt stricter protective measures. The 1991 Belarusian legislation on the management of post-accident contamination identified three categories of victim, which covered around two million people: 'liquidators' (116,000 people), 'persons evacuated in 1986' (135,000 people), 'persons living in contaminated zones' (1.6 million people).[26] It also categorized the contaminated territories into five zones: the exclusion zone (located in the 30 km perimeter), the zone of obligatory and immediate relocation (where individual doses exceed 5 mSv and the deposition of Caesium is greater than 40 ci/km^2), the zone of subsequent relocation (where individual doses exceed 5 mSv and the deposition of Caesium is above 15–40 ci/km^2), the zone of voluntary housing (where the individual doses are between 1 mSv and 5 mSv and the deposition of Caesium is below 15 ci/km^2), and the zone of periodic radiological checking (where the individual doses are less than 1 mSv – this zone was excluded from the specific interventions of the state). For the zones of subsequent relocation, of voluntary relocation and of periodic radiological checking, restrictions for food and for economic and agricultural activities were set. Financial and social compensations, such as health monitoring, additional salary, supplementary holiday, early retirement, food vouchers and visits to sanatoria were also put into place.

In particular, the creation of the zone of voluntary relocation illustrated a major change in post-accident management. This category, which affected one sixth of the people living in the contaminated zones, implied that the state left the decision whether or not to stay in the contaminated regions to the individuals themselves, thus partially renouncing its traditional role as manager of public health.[27] In other words, it was no longer the state which would determine the suitability and the cleanliness of living areas. The state's responsibility was alleviated whilst the individuals were given 'freedom' to decide themselves whether or not to stay in their homelands. But what does the 'freedom of choice' mean in a context of greater economic insecurity? Indeed, the voluntary character of relocation for people considered as moderate victims would soon turn out to be a delusion in many cases, since the decision to stay in the contaminated territories would mostly depend on the financial resources of the inhabitants, allowing them to move elsewhere, or not.

Even though the relocation plans excluded a substantial proportion of the victims of the contaminated zones, the system of protection and compensation established in 1991 implied exorbitant costs for the ex-Soviet republics. In 1991–2, expenses for compensation and relocation amounted to 22.5 per cent of the national budget in Belarus, and to 16.5 per cent of the national revenues in Ukraine. As a result of the difficult economic conditions of these countries, five years after the setting up of the relocation plans, it proved to be impossible to implement them completely. Most of the time, the relocated people simply moved to regions less contaminated then their own, but still very contaminated. Furthermore, many of the newly constructed buildings intended to re-house victims were abandoned without being used at all, since they turned out to be contaminated as well.[28] In 1994, approximately 42,000 people were still living in a zone of 'subsequent relocation'.[29] Six years later, this number stabilized at 28,000.[30] Since the state could not provide real solutions, thousands of people were condemned to stay in places that were officially considered as inappropriate for long-term accommodation.

Given these aspects, ten years after the disaster, the relocation ceased to be a priority for the Belarusian authorities. Starting from 1996, the government introduced a 'new concept' which promoted a definite return to normal in contaminated territories.[31] After ten years of health monitoring, research and humanitarian aid, it aimed to turn the page and look towards the future, end compensation measures as well as the health monitoring of risk groups, and abandon the evacuation and relocation programme of 1991.

The abandonment of the relocation policy was recommended by international bodies from the end of the 1980s. The above-mentioned IAEA report on p. 139 reflects the extent to which decisions related to evacuations and relocation were (and are) crucial for the Western world. In 1991, the IAEA was insisting on the monetary costs and the psycho-sociological impacts of relocation, such as anxiety and social distortion.[32] Additional evacuation plans were not recommended at all, as relocation was said to engender real psychological problems.[33] Given the fact that, in 1991, 200,000 new evacuations were envisaged in the contaminated territories, above all, the future of the nuclear industry was at stake. Carrying out additional evacuations several years after the disaster would aggravate the costs of a nuclear accident in economic, political and symbolic terms. It would discourage future investors and further ruin the nuclear industry's image, reinforcing the hostility of the European population *vis-à-vis* nuclear energy. The 1991 report of IAEA thus affirmed the necessity of studying the factors allowing relocation to be minimized. Five years later, the IAEA experts adopted a similar policy and put an accent on the 'trauma of relocation'.[34] A report from the Nuclear Energy Agency, published the same year and updated in 2002, insisted on the psychological effects of relocation, and even of other compensation measures.[35] Many psychological and sociological surveys were launched within this frame, with the aim of analysing the conditions that could make the normalization of contaminated territories possible.

One of them is the Commission Programme for the Evaluation of the Consequences of the Chernobyl Accident (1991–5), aiming to investigate the efficiency of the protective measures taken in the contaminated territories. The project was commissioned to two French experts, one of them the director of a private consultancy firm (Mutadis) specializing in the governance of risky technologies, the other one a sociologist-psychoanalyst at Caen University. The surveys conducted by these experts with several hundred people from Ukraine and Belarus revealed a high-level of anxiety *vis-à-vis* radioactive contamination as well as a lack of trust in the authorities, a feeling of rupture, insecurity, powerlessness, hopelessness due to an irreversible situation – 'nothing will be the same again'. These studies also indicated that the number of inhabitants demanding relocation had diminished gradually. The International Chernobyl Project had revealed a very strong public demand for relocation, with 83 per cent of the interviewees asking for it.[36] The surveys conducted in Ukraine between 1992 and 1993 showed that such a demand still existed but that trust in authorities declined, extinguishing hopes of an exit from contaminated zones.[37] In a similar manner, the field observations made in Belarus in 1994 revealed a loss of hope. The atmosphere of protest had faded away; concern for health had lost its urgency because of the abundance of socio-economic difficulties; the public had learnt to 'normalize' and 'relativize' radioactive contamination ('we got accustomed'); the idea of staying in the contaminated zone forever was becoming acceptable because of the slowness of official initiatives.[38] These observations impelled the Mutadis experts to adopt a new approach to guide people who did not want to or who could not afford to leave their villages to restart their lives elsewhere. In particular, the 'voluntary relocation' turned out to be an important 'laboratory zone' to envisage the rehabilitation of contaminated territories, given the fact that there was no formal definition of unacceptable risk in these zones, and that the social construction of an acceptable contamination seemed to be possible. The Ethos project was launched in this context in 1996. In ancient Greek, 'ethos' signifies 'custom', 'habit'. Indeed, the adoption of new habits, appropriate for living in a contaminated world, would rapidly become a key feature of the Ethos project.

The Ethos Project: From Victims to the 'Empowered'

The first phase of the project, called Ethos 1 (1996–8), was carried out in Olmany, a Belarusian village with a population of 1,300, situated 250 km from the Chernobyl reactor, in the district of Stolyn, one of the twenty Belarusian districts contaminated by the accident. Olmany is an agricultural village whose basic local products are milk, wheat and meat. This village was chosen for the deployment of the Ethos project precisely because it is part of the 'voluntary relocation zone'.

Financed by the European Commission, the coordination of the Ethos project was delegated to Mutadis and to the Nuclear Protection Evaluation Centre (CEPN), a think tank on radioprotection created with an NGO status in 1976 through an initiative of the French nuclear industry. The Ethos project assembled an interdisciplinary team (radioprotection experts, economists, sociologists, agronomists, specialists in public communication) consisting of the specialists of CEPN and Mutadis as well as academic staff from the National Institute of Agronomy of Paris-Grignon and from the University of Compiègne. Partnerships were established with local authorities, particularly with the Belarusian Ministry for Emergencies and the Belarusian Chernobyl Committee, charged with managing the consequences of the disaster. The Ethos project was conceived as a tool from Europe for modernizing the management of the consequences of the accident by the Belarusian state. From 1996 to 1998, the French team carried out twelve missions, each with a duration of ten days.

The Ethos project promoted a very strong engagement by the local populations in the rehabilitation of the contaminated territories. The term rehabilitation referred, in this context, to a joint promotion of public health protection and socio-economic advancement, with reference to sustainable development.[39] Under the banner of 'practical radiological culture', various participatory projects were set up to allow people of the contaminated territories to improve their living conditions and to take charge of their own lives. The mothers, for instance, were taught how to measure and control the radiological exposure of their children. A project was launched for farmers to improve the 'radiological quality' of 'sensitive products', in other words, the 'radioactivity level' of the 'contaminated foodstuffs'. The teachers in the village also joined the Ethos project, with the aim of integrating the 'practical radiological culture' in school programmes and developing a new pedagogy on the so-called 'radiological situation' of the village – an expression which henceforth replaced the term 'contamination'. Indeed, vast series of concepts and linguistic tools were developed in order to facilitate the construction of a new life in the contaminated territories.

The conviction of the Ethos team was that the nostalgia for the good old days before the disaster should henceforth be over. The Ethos project aimed, above all, to be a project of hope allowing the victims to live and to behave in a different manner. The victims of Chernobyl were called upon to combat their fatalism, regain confidence in the future and in themselves, and become autonomous. But paradoxically, the autonomy of individuals was called upon in a context of great heteronomy. A general definition attributed to the term autonomy is 'the quality or state of being independent, free and self-directing'. On the contrary, people of Olmany were deeply 'heteronomous', which means 'subject to external controls and impositions', given the fact that radioactive contamination dictated the principal rules of action and imposed the most important restrictions on them in their interaction with the environment and also with their own bodies.

Autonomy did not only serve as a rhetorical tool or a 'discursive regime'.[40] It was put into practice via the 'practical radiological culture' which implied not only the acquisition of a technical proficiency (basic notions in radioprotection, measuring of radioactivity, and so forth) but also the adoption of good practices to cope with an omnipresent contamination. Indeed, the aim was to generate a totally new geographical and material culture by redefining places, foodstuffs and daily life objects in relation to their radioactivity levels. For instance, the forests, which are abundant in Belarus, were no longer places reserved for long walks or relaxation. Because of the high concentration of radioactivity in these zones, they were henceforth to be considered as 'expensive' places in terms of radiation, and the time passed there was to be reduced. The inhabitants were supposed to detect the variation in the contamination from one place to another using a radiometer, in order to determine the appropriate actions to take on a daily basis. A dozen contamination maps for different parts of the village were thus prepared. The radiometer was intended to serve as a specific instrument to guarantee the 'good behaviour' of all by confronting them with the Chernobyl 'reality', by reminding them that contamination was henceforth a major component of their lives, their foodstuffs, their own bodies.[41]

Three categories of places were designated with reference to the average dose of exposure to medical radioactive sources in France (i.e. 1.3 mSv/year). The first category characterized the places which were not very contaminated (with an exposure level below the reference value) and thus could be visited; the second category concerned places where the duration of exposure had to be reduced (between 1.3 mSv/year and 8.7 mSv/year); the third category concerned places to avoid because of their high level of contamination (above 8.7 mSv/year). Within this framework, the houses were said to be safe places (category 1); most of the private gardens became places to visit less frequently (category 2); the forests, hay ricks or the proximity of domestic stoves (often polluted by the ashes of the forest wood) were the least recommended (category 3). The important point here is that visiting these places was not banned but was optimized, in terms of risk, through the promotion of good practices. This is what the Ethos team called the 'Alara attitude', with reference to the Alara (as low as reasonably achievable) principle of optimization in radioprotection. The individuals would henceforth spend less time in their gardens, or at least compensate for the time passed there by staying longer in the home. Indeed, the optimization plan was to be applied even within houses as certain families realized, thanks to measurements taken at home, that the contamination level of their houses was not homogeneous and that certain rooms seemed more contaminated than others. By adopting the 'Alara attitude', the families were urged to minimize the time passed in contaminated rooms or wash their houses more frequently in order to prevent the migration of radioactive dusts from polluted rooms to those considered clean.

The foodstuffs were also redefined and re-categorized in terms of their radioactive charge. Following a vast measurement campaign, all food products consumed in the village were divided into three categories according to their levels of contamination : (1) products with a radioactive contamination higher than the maximum authorized doses (mushrooms, berries, pork, fish, milk products); (2) products with a moderate contamination (beetroot, potato, carrot, cucumber, cabbage); (3) 'neutral' products (i.e. those imported from abroad or 'clean' regions and/or those controlled by the state). Thanks to such a categorization, mothers, for instance, would make up the diet of their children by calculating the 'radiological contribution' of each foodstuff (an expression that replaced 'radioactive contamination') and adjusting different aliments in order to maintain a low exposure to radiation. A system that can be qualified 'food rationing by dose' was thus established, the rationing being no longer determined by the scarcity of foods but rather by their richness in radio elements. Indeed, the notion of 'annual intake budget' was developed to facilitate the management of food rationing. The ideal 'annual budget' was fixed at 20,000 Bq with a recommended upper limit of 100,000 Bq (1.3 mSv per year).[42] According to the Ethos team's annual intake budget system, contaminated products, especially those from category 1, became consumable, but in moderation, since they were 'expensive' in becquerels.

Prior to the Ethos project, despite the official prohibition, most of the contaminated foodstuffs were already being consumed by a substantial part of the local population. That was the case in particular for the basic foods (milk, meat, fish) which are difficult to exclude from a daily diet but are also impossible to replace with highly expensive imported products. Many villagers were therefore forced to ignore the toxicity of these products, and, in so doing, they felt guilty *vis-à-vis* their children. The 'practical radiological culture' of Ethos proposed the consumption of these products in full knowledge of the situation, in a rational and optimized manner, without feeling humiliated or guilty. For the Ethos team, getting rid of prohibitions was a major condition for improving the 'quality of life' of the inhabitants who had hitherto been subject to numerous restrictions. This was also considered a good way to help people accept that they were henceforth living in a 'contaminated universe'.

In this new and irreversible universe, the norms of radioprotection and the limit doses no longer had any significance. A norm serves to distinguish the good from the bad, the acceptable from the unacceptable, and the normal from the exceptional. Establishing a norm necessarily means restricting certain practices. According to the Ethos framework, the management of risks by norms was no longer functional even though the norms legally existed and orientated public action in a general manner.[43] Locally, the notion of norm was replaced by the notion of 'reference value' which had no regulatory status and which was simply a guidance tool intended to facilitate individuals' actions. In Olmany, with the Ethos project, the notion of dose also became obsolete and was replaced by the notion of 'annual intake budget'. Like the 'reference value', this notion was merely an informal indicator, an advisory tool to guide individual risk-taking. The 'empowerment' of individuals gained a central importance within this process as a politics exported from Europe. Vigilance, food rationing, redefinition of geographies and products, but also self-restraint – it was essential to overcome the temptation to walk in the woods or to eat berries of the forest, for instance – became the new reflexes, virtues and customs that were supposed to permit the empowerment of local populations.

The politics of empowerment of the villagers of Olmany was not limited to the adoption of a good consumer attitude in a new context. As local producers, the villagers were also invited to revise their farming practices in order to diminish the contamination level of their products and, if possible, to reintroduce them into the market. When the Ethos project was launched, the commercialization of milk and meat (i.e. the two major products of Olmany) was forbidden. Because of the high toxicity of the hay and the grazing land, the milk was the most contaminated product of the village, with a radioactivity level twenty times higher than the authorized limits. Since the hay was expensive, just like ferrocene (an additive which helps reduce the transfer of Ce137 from the fodder to the milk of the cows), the Ethos project urged the cattle raisers to optimize the food of their animals in both radiological and economic terms. In this context, the utilization of the most contaminated hay was envisaged during the period of gestation or only for the young heifers that do not give milk. The feeding of the cattle with less contaminated hay was prescribed only during their milking period. The utilization of ferrocene was also indicated during this period, and especially when the fodder was very contaminated. Thanks to this approach, the 'radiological quality'– an euphemistic expression employed by the Ethos team – of milk from Olmany was improved to such an extent that it was re-commercialized in March 1998.

The 'empowerment' of the public in the contaminated territories was thus closely correlated with the new conditions imposed by a nuclear accident. Individuals had to learn how to survive in an intractable environment, by transforming the contaminated lands into cultivated ones. The Ethos project thereby gave shape to the idea that the consequences of a nuclear catastrophe should be managed by individuals themselves who are expected to become informed, proficient, autonomous and 'responsible'. A new, neo-liberal 'art of government' was formulated with a major objective: to delegate risk management to the private sphere. Within this framework, the empowerment served as a major technology for governing people.

It corresponded to what Emilie Hache called 'liberating responsibilization', which is a specific and positive (or, positively connoted) form of transfer of responsibilities towards individuals, by urging them to interiorize their social situation, to be concerned about themselves and to manage their lives on their own.[44]

Indeed, both in technical and economic terms, the post-accident contamination observed in Belarus turned out to be 'impossible to manage' solely with state experts and organs. As we have seen above, it proved too expensive to relocate the inhabitants of the contaminated zones, all the more so because relatively clean zones seemed to be very rare in Belarus. And when the local populations stayed in the contaminated zones, it became impossible for the state and its experts to really manage the situation, given the complexity of radioactive contamination. The lifestyle of individuals, their diets or the time they passed in the forest every week had a direct influence on the dangerousness of the exposures. Preventing risks in these conditions was not possible unless the state precluded access to most places and banned most of the foodstuffs. Since all this was difficult to achieve in the case of Belarus, the inhabitants of the contaminated zones – the 'victims' – were transformed into 'experts'. The participatory governance promoted in this frame was called 'co-expertise', which signified that there was no available expert advice to manage the 'complex situation' (a term which replaced 'crisis') of the contaminated territories, and that knowledge should be built up jointly with the victims.[45] In other words, public participation in risk management turned out to be the only way to confront a permanent contamination, to minimize relocation and other measures, to prevent health scandals and thus to 'normalize' the post-catastrophe period. During our interview, one of the coordinators of the Ethos project explained this change in the following manner:

> Chernobyl revealed a new category of public: several millions of people living with radioactivity on a daily basis. Nobody had envisaged it. Prior to Chernobyl, each case [of contamination] was manageable by professionals, by radioprotection specialists. In Chernobyl, you cannot assign one expert per home, per citizen. That's why there is a vast reflection. Either you say that it is very complicated and you take charge of the problems in the name of people. In this case you exclude them. Or you re-engage people in the process; you go for a co-management of the situation. The experts do their jobs as experts and the people do their best to protect themselves. If you were saying, 'From now on, regarding dental hygiene, we will not require anything from the public, only experts will take charge of it', the whole system would collapse.[46]

In this context, Olmany served as a preliminary 'laboratory'[47] to explore channels of survival in a contaminated world.However, the experts of the Ethos project refused to evaluate the acceptability or not of such a world, despite the fact that this question was of direct concern for the inhabitants of Olmany. During the first public meeting organized by the Ethos team, most villagers tried to ascertain whether, according to the European experts, it was appropriate to live in the village, and whether these experts would let their own children live there. The Ethos team left these questions open, arguing that the decision to live in the village did not belong to experts but to the public. They clearly defined their role as that of helping to improve the living conditions of those deciding to stay there.[48]

The Ethos project managed to attract a substantial portion of the villagers. It gained credibility thanks to the concrete aims it identified (e.g. improving the quality of milk), and the 'active dialogue' it proposed with the local population. According to the Ethos activity reports, between July 1997 and December 1998, the body contamination of children who followed an optimized diet diminished in a significant manner, whereas, till then, the only way to improve children's health had been to send them to sanatoria abroad during holidays. Inevitably, such concrete improvements provided an implicit answer to the fundamental question of 'can we really live here?' During interviews carried out by the Ethos team,[49] an inhabitant of Olmany explained how he perceived the Ethos mission: 'The French Ethos group came here to investigate solutions to contamination. It seems that solutions exist. On the lands of Olmany, we can grow and eat clean products.' Another citizen indicated, in a similar manner, that 'the Ethos project shows that we can take action ourselves. It is a simple programme. If we can manage our health ourselves, then we can stay in the village.'

Rapidly, the Ethos approach attracted the attention of the local and national authorities of Belarus who expressed their willingness for the implementation of the project in the whole district of Stolny. It also gained the support of the French nuclear industry (EDF, Cogema), the French public authorities (IPSN– the Institute for Radioprotection and Nuclear Safety), the Swiss government and the European Commission who decided to finance a second mission. In this second phase, the project was expanded to five villages of Stolyn and addressed almost 10,000 inhabitants. With the aim of transferring the French expertise to local specialists, about eighty professionals (nurses, doctors, teachers) were recruited on a voluntary basis.

The second Ethos project was completed in 2001 and concluded with an international seminar organized in Stolyn with the participation of 150 people from several Belarusian, French (EDF, IPSN, universities) and international (European Commission, UNDP, World Bank) organizations. A common declaration, adopted during this event, insisted on the necessity to support the Ethos approach to the rehabilitation of the contaminated territories. During the same period, the Belarusian government decided to reduce its relocation programme and social assistance. Just after the Stolyn conference, 146 villages were removed from the list of contaminated zones, a decision that deprived around 66,000 people of social help, healthy food and regular visits to sanatoria in Russia.[50]

Participatory Rehabilitation: From a Local Experiment to an International Reference

The Ethos approach and its orientation towards a 'participatory' and 'sustainable' rehabilitation had an important echo in the international arena. In 2002, a joint UNDP–UNICEF report on human consequences of the Chernobyl accident put, for the first time, an accent on the socio-economic impacts of the disaster (beyond its health impacts) and recommended the reinforcement of projects like Ethos.[51] A report from the World Bank, published the same year, indicated, in a similar perspective, that

> economic development should be a key priority of the [Belarusian] Government to shift from programmes that create a victim and dependency mentality to those that support opportunity, promote local initiatives, involve the people and spur their confidence in shaping their destiny.[52]

The victim and dependency mentality mentioned here referred to individuals from socialist regimes taken care of by the state (through the USSR law of 1991, for instance). The World Bank rejected such a stereotype, considered archaic and old fashioned. It promoted 'autonomous' and 'entrepreneurial' individuals, who were supposed to be adaptable to the neo-liberal system. In a period of transition to the market economy, the Belarusian state was thus recommended to encourage the emergence of such new, 'modern' individuals. This signified the minimization or the removal of aid and compensation, or even the abolition of the legal status of victim. The World Bank had already recommended Ukraine adopt such a policy a few years before.[53]

In 2003, the Ethos experience led to the launch of a vast international programme called Core (Cooperation for the Rehabilitation of Life Conditions in the Contaminated Territories of Belarus). With the aim of making the Ethos approach the reference in post-accident management, the Core programme extended the project to four districts (Stolyn, Bragin, Cherchersk and Slavgorod) situated in the three most contaminated regions of Belarus. Coordinated by the creators of the Ethos project, it was financed by the European Commission, the French, Belarusian and Swiss governments, and a few UN agencies (UNDP, Unesco, Unicef and the United Nations Fund for Population Activities), with a budget of EUR 8 million in 2005. The Core programme was envisaged as a vast network mobilizing hundreds of actors: Belarusian and international agencies, European foundations, universities, research centres and journalists as well as Belarusian, European and American NGOs. Western NGOs engaged in humanitarian aid, radioactivity measurements or cultural and pedagogical activities were thus integrated in the promotion of participatory governance in the contaminated territories. However, they were called upon as the kingpins of the project, whereas the high-level decisions were reserved for official experts and diplomats such as, for example, the ambassadors of nine European countries in Belarus.

The declaration of the principles of the Core project promoted the 'engagement of the civil society' as an 'innovative approach' to improving the living conditions of the victims of Chernobyl, who henceforth were to be considered 'partners'.[54] Within this framework, key notions such as active participation by local actors, sustainable rehabilitation or adaptive management of a complex situation entered into the vocabulary of post-accident managers. Furthermore, with the Core programme, disaster talk ceased to be a taboo. On the contrary, it was encouraged via cultural and pedagogical projects aiming at an 'intergenerational and international transmission of the memory' of the Chernobyl disaster. One such project, entitled *Tell me about the cloud*, showed documentaries on the Chernobyl cloud not only in Belarus and in Europe but also in Africa, twenty years after the accident. Such efforts were intended not only to open up a new 'era of transparency' in the nuclear domain but also to 'condition' the people to the idea of the unavoidability of the 'risk society' and of long term radioactive contamination.

Chernobyl: A 'Sage' Lesson for Europe?

One major motivation behind the Ethos project and the Core programme was that they were conceived as a 'transfer of experience' (*retour d'expérience*) for Europe. Long-term post-accident management and rehabilitation (beyond the emergency phase) became a major political concern in France and in Europe after the tragic events of 11 September 2001 which put the terrorist risk on the agenda. The promotion of a 'nuclear renaissance' as an answer to the climate change problem and relevant plans to double or even triple the number of nuclear power stations worldwide also played a role in the rise of political interest in post-accident management, since the prevention of malevolent action on nuclear installations was considered more difficult with a greater number of reactors.

In 2002, prior to the start-up of the Core programme, a European project had been launched to develop a practical radiological culture in case of a severe nuclear accident in Europe. The so-called 'Sage' project aimed to produce a practical guide for the actions of the European people in case of long-term contamination caused by a nuclear accident. This was an innovative attempt given the fact that, until then, the action plans had dealt with only short-term/emergency management.[55]

The Sage project largely leant on the Belarusian experience and the Ethos approach, especially with regards to the central role assigned to local populations in the daily management of radioactive contamination. It was implemented by experts from French (CEPN), British (National Radiological Protection Board) and German (National Research Centre for Environment and Health) organizations as well as two Belarusian radioprotection institutes. With the Sage project, for the first time Europe – and no longer Belarus – was the focus of attention. The probability of a serious nuclear accident in one of the European nuclear installations was officially accepted and was even opened for discussion. The Sage project was promoted as a participatory tool, as a means of offering the 'actors' the possibility of managing their (own) future.[56] To produce the Sage guide, stakeholder panels were set up in three European countries (France, England, Germany) with the participation of NGO members as well as professionals coming from different horizons (doctors, nurses, teachers, farmers).

Comprising fifty pages and six chapters, the Sage guide gave instructions on the utilization and the interpretation of radioactivity-measuring instruments.[57] It also described the mission of health professionals in the case of a very long-term contamination. However, based on the Stolny experience, the Sage guide mainly concerned the management of rural regions. Moreover, it took into consideration Caesium-137 only, thus ignoring the problems related to the concentration of many other radionuclides in water, soil or the human body. Clearly, the basic aim of such a guide was not so much to take control of a hypothetical post-accident situation but rather to 'raise public awareness of new risks related to radioactivity'.[58] Indeed, as the name attributed to the project suggested (the two widespread meanings of the term 'sage' in French are 'wise' and 'well behaved'), the Sage guide was above all a symbolic tool with a double function. The first was 'pedagogy on disaster', the idea being to accustom people to the reality of a potential risk which might change their living conditions in an irreversible manner. The second was 'pedagogy on responsibility', which consisted of stimulating the sense of responsibility of individuals 'before' the disaster took place, so that they could behave 'correctly' in the event of a disaster.

Through a modern guide endowed with various virtues (it was prepared in a pluralist manner, with the participation of different stakeholders; it reflected a will to inform citizens, to act in a transparent way, to promote participatory democracy), individuals were, thereby, invited to become aware of the active role that they were supposed to play in a post-accident situation. Another original feature of the Sage project was that, for the first time (in the nuclear domain at least), actors were called upon to participate in the resolution of a problem which did not (yet) exist. The issue was the co-construction of the management of a contaminated future whilst the ways to avoid such a future were not open to discussion. The participatory approach adopted by the Sage project had an additional advantage for the authorities: it gave an opportunity to test, in advance, the reactions of the public *vis-à-vis* the eventuality of living in a contaminated environment someday, just like the two million Belarusian citizens.

Is France Preparing Itself for an Accident like Chernobyl?

The Sage project was followed by other projects, especially in France. In 2005, the French Nuclear Safety Authority commissioned the Ethos experts (Mutadis and CEPN) to develop a long-term management strategy in case of an accident or a terrorist attack in the French nuclear park. The so-called Parex (*Post-Accidentel Retour d'Expérience*) project, conceived as a participatory arena, assembled experts from ministries and health agencies, industrialists, local governors as well as NGOs. Two major lessons from the Belarusian experience were followed. The first was that, in the event of a severe accident, outside the exclusion zone managed by the state, individuals should have the freedom to decide whether they would stay in a contaminated region or leave it. In other words, the Ethos experts suggested, implicitly, that in the case of a nuclear disaster in France, most of the contaminated zones be transformed into 'voluntary relocation zones' (with reference to the Belarusian categorization of 1991). The second lesson from Belarus, within the Parex project, concerned the necessity to allow local populations to take initiatives on their environment, to facilitate the shift from a logic based on the respect of norms to a logic of establishment of the 'quality of life' for individuals and families.[59]

The Parex project served as a preliminary step for a vast governmental action launched a few months later. In June 2005, the French Nuclear Safety Authority set up a committee, the so-called Codirpa, charged with developing a post-accident doctrine for France. Codirpa was composed of delegates from the Nuclear Safety Authority, the Institute for Radio protection and Nuclear Safety, the General Secretary of National Defence, as well as ministries (Agriculture, Budget, Defence, Ecology, Health, Industry and Interior), health agencies and NGOs. It is noteworthy that the nuclear industry was excluded from the committee as the long-term management of a nuclear accident was considered a joint mission for public authorities and 'civil society'. A Codirpa member explained during our interview: 'Given the fact that nuclear operators cannot be held responsible for a nuclear accident caused by a hostile action, they are not invited.' He insisted: 'What can we expect from them in the long term anyway? They will probably have disappeared [after a severe accident].'[60]

During its initial phase, Codirpa dealt with relatively moderate nuclear accidents (those that would cause neither definitive evacuations nor prolonged radioactive releases), setting aside major accidents resulting in releases of plutonium and/or in the fusion of the reactor.[61] Around ten working groups were constituted to study a large range of complex problems such as the cessation of emergency actions, the monitoring of public health, the rehabilitation of rural areas, the compensation of victims, the coordination between public authorities and stakeholders, the management of radioactive waste, and so forth. Several

NGOs, such as GSIEN (Group of Scientists for Information on Nuclear Energy, founded in 1976 and much inspired by the American Union of Concerned Scientists), ACRO (Association for the Control of Radioactivity in the West, created in 1986 to provide independent information and expertise in the nuclear domain), Mères en colère ('Angry Mothers' – a mothers' collective founded in 1997 during a strong scientific controversy on child leukaemia around the Hague reprocessing centre), were integrated in the working groups.

To summarize, a very significant change took place in post-accident management during the last decade. While in the 1970s the authorities affirmed publicly that a severe nuclear accident was almost impossible in France and the fundamental documents related to emergency management were kept secret, three decades later, actors critical of nuclear energy are invited to negotiate the co-management of the consequences of a possible nuclear accident in France and in Europe. The operators are excluded from the committee whereas the 'critical vision' of NGOs is called upon for the resolution of a problem laden with uncertainty: what does it mean, indeed, to model and to plan the long-term management of the radioactive contamination of a whole city or even a whole country? Within Codirpa, the NGO members are also assigned the role of mediator between the government and the public, facilitating the transmission of a new message, a new discourse already delivered to the media arena, as illustrated by the title of an article published in the daily newspaper *Le Monde* (21 February 2008): 'France is preparing itself for the consequences of an accident like Chernobyl.'

From the mid-1990s, the public authorities admitted publicly the risk of a serious accident in the French nuclear park. The integration of accident simulation exercises in the school syllabus during 1995 as well as the decision taken in 1996 to distribute stable iodine prophylaxis to people living close to a nuclear installation (within a 5 km perimeter) are representative of such a development. Nevertheless, during this period, the aim was still to minimize the immediate consequences of an accident. The possibility of the contamination of vast areas several thousand kilometers away from the accident area, or of a long-term pollution lasting several decades, even thousands of years, was not pronounced publicly. The radically new character of the Parex approach and the Codirpa programme is that the authorities have started to publicize such discourses. Some Codirpa experts even affirm that our societies are in a 'pre-accident' phase. It is thus suggested that we have already taken the road towards the disaster, and that everyone should, from now on, establish continuity between the pre- and the post-accident contexts. Nevertheless, for the moment, if the above statements are expressed within 'participatory' arenas like Codirpa, they are not propagated more widely. Furthermore, the French government seems only to consider modest accident scenarios instead of dealing with the problem in all its aspects. It seems difficult, for instance, to include a scenario about an accident happening in the Hague reprocessing plant in Codirpa's agenda.[62]

A Contested 'Future'

A rapid glance at the emergence of a variety of new projects and programmes in Belarus, in France and in Europe since the second half of the 1990s demonstrates the way in which participation of 'stakeholders' has gradually gained importance in post-accident management. It has since then been admitted that, in view of the Belarus example, it would be *impossible* for states to take total charge of the consequences of a long-term contamination. Nevertheless, civil society organizations engaged in these projects are quite reduced in number. In the French context, the NGO stakeholders invited to take part in projects like Core, Sage or Parex or within the Codirpa programme, are almost always the same. In particular ACRO and GSIEN, both of whom act as counter-experts in the nuclear domain,[63] have agreed to join these participatory bodies, together with a limited number of associations concerned with environmental protection or consumer rights but mostly invisible in nuclear protestation. Indeed, the Ethos approach did acquire real support in the institutional arena, but not necessarily among the protest groups. On the contrary, projects based on the Ethos approach generated a strong rejection, especially in France.

In 2005, an international Sage seminar organized in Conservatoire National des Arts et Métiers (14–15 March 2005, Paris) suffered a spectacular sabotage. A couple of masked activists entered the conference hall and threw rotten eggs and red ink ('fake blood') at the organizers of the seminar as well as the foreign participants (Belarus, Germany, England, Belgium, Norway). The 'anti-Sage commandos', as the press called them, accused the Sage project of trivializing post-accident contamination and of pretending that one can live normally in contaminated zones with the benefit of a 'guide indicating how to die while counting becquerels'.

In 2006, a book entitled *Crime de Tchernobyl: le goulag nucléaire* (Chernobyl Crime: the Nuclear Gulag), written by the film producer Wladimir Tchertkoff, co-founder of the association 'Children of Chernobyl-Belarus' (created in 2001 to help the Belarusian children suffering from the disaster), heavily criticized the Ethos project and the Core programme. It denounced the neglect of public health within these projects. It accused them of merely favouring economic development under the banner of sustainable development, as an extract from the book indicates: 'Before "sustainable development", children were happy in this splendid nature. Today, they are gloomy, depressed, weakened.'[64]

The Ethos project was first criticized in 2001, when the Belarus government withdrew the management of the local radiological control centres from the

Belrad Institute, an independent Belarusian institute for radiological protection that had set them up in five districts of Stolyn ten years before, with the object of allowing the local populations to measure the contamination of the local products. Within a context of repression *vis-à-vis* the scientists contradicting the official thesis of a return to the normal conditions in the contaminated territories,[65] Belrad's action became undesirable, all the more so because Vassili Nesterenko, director of the Belarus Institute of Nuclear Energy from 1977 to 1987, contradicted the official policy by demanding more precautionary measures. In particular the measurements made by the Belrad Institute on the body contamination of around 200,000 children through mobile laboratories equipped with anthropo-gamma-meters proved to be troublesome in official terms. The research conducted by V. Nesterenko showed that the proportion of unhealthy children rose from 15 per cent before the Chernobyl accident to 80 per cent (even 94 per cent for the Gomel region) after the disaster.[66]

For the Belarus government, Belrad's action, mainly centred on public health, represented the *ancien regime* whereas the Ethos approach mobilized quite a different argument advocating socio-economic development as well as less involvement of the state. Hence, at the local level, many tensions occurred between the Ethos team's 'sustainable rehabilitation' project and Belrad's health programme. In 2001, Professor Nesterenko expressed doubts on whether the Ethos approach really permitted diminishing the body contamination of children.[67] Michel Fernex, an anti-nuclear doctor and a friend of Nesterenko, criticized the coordinators of the Ethos project, especially CEPN, for acting as a covert organ of the 'nuclear lobby' and controlling the dissemination of information on children's health.[68] The CRIIRAD (Commission for Independent Information and Research on Radioactivity), an NGO aiming to provide counter-expertise in the nuclear domain since 1986, accused CEPN of channelling public and European funds to the nuclear lobby and thus preventing 'genuine' independent and helpful projects in Belarus.[69]

The Ethos approach also provoked a division within the NGO arena where all three types of public commitment described by A. Hirschmann (exit, voice, loyalty) were mobilized.[70] A split occurred between those who promote an 'idealist' position, claiming that it is not acceptable to let the victims live in the contaminated zones, and those who hold a 'pragmatic' position, considering that the victims should not be abandoned to their own destiny in such difficult circumstances, under the pretext that they cannot be re-housed. A similar dilemma divided those participating in committees like Cordirpa and those expressing an 'ironic detachment from technical democracy'[71] considering that it is not possible to participate in the management of (nuclear) problems, given the fact that the state had never consulted the public over the construction of the nuclear power stations which *generate* these problems. For instance, ACRO took part in the Core programme as well as the Parex, Sage and Codirpa projects. ACRO considered that 'although it is not normal to live with such contamination levels, people do live there and they do need help.'[72] The association also judged that joining the Core programme was the only way to collect data on the 'disastrous situation' of Belarus. The choice of ACRO was nevertheless considered as being collaboration with the nuclear lobby by CRIIRAD and by several anti-nuclear groups. In 2008, at an international conference entitled 'Living after Chernobyl' (Lyon, 15–16 May 2008), the Core programme was heavily criticized. A member of ACRO explained:

> People from Réseau Sortir du Nucléaire [the French Network for Phasing Out Nuclear, bringing together around 800 organizations] attacked me [for joining Core]. They then turned to the Belarusian NGOs and told them: 'Listen, we have nothing against you. But please be aware that you have become a laboratory. They make use of you as a laboratory.' An ex-inhabitant of Pripyat immediately stood up and said: 'We know that we have become a laboratory, an open laboratory. So what? You should also come and work with us.' That produced a chill in the room.[73]

Concluding Remarks

Recent policies of rehabilitation in territories contaminated by the Chernobyl accident reveal four major issues:

(1) The *empowerment* of the public in the context of long-term radioactive contamination corresponds to a new way of governing people who should learn how to *survive* on their own, by measuring radioactivity outside and in their homes, by optimizing the radioactive content of their foodstuffs, by calculating the diet of their children on a daily basis, by adopting permanent vigilance, autonomy and self-control as new forms of self-governance. Empowerment thus appears to be a specific tool for educating people. It is a process of convincing individuals that they are henceforth living in an irreversibly contaminated world, that the *good old days* will never come back, and that the only 'solution' is to accept the new reality (i.e. life with radioactivity) and to take initiatives instead of being fatalistic and expecting the state to save its citizens. In Foucauldian terms, individuals are given freedom in a predetermined context so that their actions can be better orientated.[74] The invention of new linguistic codes reflecting a 'happy' governance (e.g. the term 'radiological quality' replaced 'radioactive contamination') is judged crucial for realizing such changes, and for establishing a new world order with new normalities.

(2) Generally speaking, the politics of empowerment, exported from France by the Ethos team, is incontestable as such, given the fact that the victims expect urgent improvements in their life conditions and that their biggest fear is to be abandoned and forgotten. I argue, however, that paradoxically, the promotion of empowerment in the context of long-term radioactive contamination may imply

a new form of abandonment of individuals by the state and the international bodies. In post-accident management, far from being a step towards the democratization of science, the politics of empowerment corresponds to the transfer of responsibilities from the state to the public. In the case of a serious nuclear accident, such a transfer seems almost inevitable for two basic reasons. The first is that the radioactive risk observed in countries like Belarus proved almost impossible to be managed by state experts alone because of the technical complexity of the contamination, widely diffused in both geographical and biological terms (i.e. victimized bodies). The second reason is that, in economic terms as well, state-centred risk management in highly contaminated regions turned out to be almost impossible to implement, given the very high financial burden that it imposes on the state (i.e. relocation and other compensations). Within this framework, the empowerment of victims corresponds to their responsibilization. Beyond its efficiency as a discursive regime, the politics of empowerment is in line with the neo-liberal trends in scientific governance towards a lesser state commitment in risk and disaster management.

(3) The fact that the politics of empowerment is intended, in the long-term, for the abandonment of individuals living in contaminated territories has important ethical consequences since these people were not consulted, prior to the construction of the Chernobyl reactor, on whether they would agree to readjust their lives according to rules dictated by radioactive contamination. The politics of empowerment seems even more problematic when it is envisaged as a new doctrine for France and for Europe. Even though it is obvious that people living in contaminated territories should be helped (via projects like Ethos and Core, amongst others), it is also clear that it is not normal to live by checking the radioactivity of foodstuffs on a daily basis, and that it cannot be presented as being normal. All individuals deserve to live in a healthy environment, without having to spend their time measuring radioactivity or optimizing becquerels. Life with permanent radioactive contamination is, therefore, a total injustice that cannot and should not be generalized to the rest of the world. There are no real solutions – only partial arrangements – for the management of contaminated territories. However, this crucial point is not stressed sufficiently by the Ethos approach and the projects derived from it, which promote 'positive' and 'happy' governance by mobilizing both democratic ideals (public participation, empowerment, autonomy, freedom of choice) and humanitarian responsibility (accompanying the victimized populations). As a result, public participation turns out to be a source of important political and moral dilemmas and contradictions, thereby provoking a division among anti-nuclear groups themselves, especially in France.

(4) Finally, it can safely be argued that the politics of empowerment, which became a key component of post-accident management, does not necessarily reflect a democratization of relations between nuclear and society in more gen-

eral terms. If public participation is promoted in contexts where the disaster has already taken place, and where harm and suffering are massive, it does not have much relevance to major decisions that are determinant for the future of nuclear energy, at least in France. One emblematic example is that, after thirty years of demand by anti-nuclear groups, a 'national debate' on nuclear energy (more precisely on the European Pressurized Reactor – EPR – which is supposed to replace the present light-water power stations, during the renewal of the French nuclear park by 2015) was eventually organized in 2006 but in unacceptable conditions since the official decision to construct the first EPR had already been taken before the finalization of the 'debate'.

NOTES

Boudia and Jas, 'Introduction'

1. WWF, *76 produits toxiques dans le sang des parlementaires européens*, Posted on 21 April 2004 at http://wwf.panda.org/wwf_news/press_releases/index.cfm?uNewsID=12622&uLangID=3 [accessed 24 September 2012].
2. A. Iles, 'Measuring Chemicals in Bodies: Biomonitoring and the Global Governance of toxics and Hazardous Wastes', Paper presented at the annual meeting of the International Studies Association, Chicago, 2007; A. Daemmrich, 'Risk Frameworks and Biomonitoring: Distributed Regulation of Synthetic Chemicals in Humans', *Environmental History*, 13:4 (2008), pp. 684–94. On CDC biomonitoring programme, see the presentation of this programme at http://www.cdc.gov/biomonitoring/about.html [accessed 24 Septembre 2012] and *The Annual National Report on Human Exposure to Environmental Chemicals*, National Center for Environmental Health, Atlanta, GA. See also: K. Sexton, L. L. Needham and J. L. Pirkle, 'Human Biomonitoring of Environmental Chemicals', *American Scientist*, 92:1 (2004), pp. 38–45; C. F. Cranor, *Legally Poisoned: How the Law Puts Us at Risk from Toxicants* (Cambridge, MA: Harvard University Press, 2011).
3. Iles, 'Measuring Chemicals in Bodies'; Sexton, Needham and Pirkle, 'Human Biomonitoring of Environmental Chemicals'.
4. J. Lawson, *Staying Well in a Toxic World: Understanding Environmental Illness, Multiple Chemical Sensitivities, Chemical Injuries, and Sick Building Syndrome* (Chicago, IL: Nobble Press, 1994); P. Brown and V. Gunter, *Illness and the Environment: A Reader in Contested Medicine* (New York: New York University Press, 2000); M. Murphy, *Sick Building Syndrome and the Problem of Uncertainty: Environmental Politics, Technoscience, and Women Workers* (Durham, NC and London: Duke University Press, 2006).
5. V. Heyvaert, 'Reconceptualizing Risk Assessment', *Recueil*, 8:2 (1999), pp. 135–43. H. Selin, 'Coalition Politics and Chemicals Management in a Regulatory Ambitious Europe', *Global Environmental Politics*, 7:3 (2007), pp. 63–93.
6. J. Applegate, 'Synthesizing TSCA and REACH: Practical Principles for Chemical Regulation Reform', *Ecology Law Quarterly*, 35:4 (2008), pp. 721–70. On the history of TSCA, see the TSCA oral history project of the Chemical Heritage Foundation at http://www.chemheritage.org/research/policy-center/projects/tsca-oral-history.aspx [accessed 24 September 2012]. Historical data can be found in: J. T. Patterson, The *Dread Disease: Cancer and Modern American Culture* (Cambridge, MA: Harvard University Press, 1989); S. S. Epstein, *The Politics of Cancer* (San Francisco, CA: Sierra Club Books, 1978); S. S. Epstein, *The Politics of Cancer Revisited* (New York: East Ridge Press,

1998); R. N. Proctor, *Cancer Wars: How Politics Shapes What We Know and Don't Know about Cancer* (New York: Basic Book, 1996); D. Davies, *The Secret History of the War on Cancer* (New York: Basic Book, 2009); S. Jasanoff, *The Fifth Branch: Science Advisers as Policymakers* (Cambridge, MA: Harvard University Press, 1990); S. Jasanoff, 'Science, Politics and the Renegotiation of Expertise at EPA', *Osiris*, 7 (1992), pp. 194–217; C. F. Cranor, *Regulating Toxic Substances: A Philosophy of Science and the Law* (New York: Oxford University Press, 1993); E. Millstone and P. van Zwanenberg, 'The Evolution of Food Safety Policy-Making Institutions in the UK, EU and Codex Alimentarius', *Social Policy and Administration*, 36:6 (2001), pp. 593–609. On the comparison of European and the US regulation of chemical substances, see R. Brickman, S. Jasanoff and T. Ilgen (eds), *Controlling Chemicals: The Politics of Regulation in Europe and the United States* (Ithaca, NY and London: Cornell University Press, 1985); D. Vogel, 'Ships Passing in the Night: The Changing Politics of Risk Regulation in Europe and the United States, Robert Schuman Centre for Advanced Studies', (Brussels: EUI Working Papers RSC n°2001/16, 2001).

7. On the management of toxic substances in the eighteenth and nineteenth centuries see for instance: C. Hamlin, *A Science of Impurity: Water Analysis in Nineteenth Century Britain* (Berkeley, CA: University of California Press, 1990); S. Barles, *La ville délétère. Médecins et ingénieurs dans l'espace urbain. XVIIIe-XIXe siècle* (Seyssel: Champs Vallon, 1999); C. Bernhardt (ed.), *Environmental Problems in European Cities in the 19th and 20th Century* (Münster: Waxmam, 2004); C. Bernhardt and G. Massard-Guilbaud (eds), *Le démon moderne: la pollution dans les sociétés urbaines et industrielles d'Europe/The Modern Demon: Pollution in Urban and Industrial European Societies* (Clermont-Ferrand: Presses Universitaires Blaise-Pascal, 2002). G. Massard-Guilbaud, *Histoire de la pollution industrielle en France, 1789–1914* (Paris: Éditions de l'EHESS, 2010); T. Le Roux, *Le laboratoire des pollutions industrielles. Paris, 1770–1830* (Paris, Albin Michel, 2011); J.-B Fressoz, *L'Apocalypse joyeuse. Une histoire du risque technologique* (Paris: Le Seuil, 2012); P. Bartrip, *The Home Office and the Dangerous Trades: Regulating Occupational Disease in Victorian and Edwardian Britain* (Amsterdam and New York: Editions Rodopi, 2002); S. Mosley, *The Chimney of the World: A History of Smoke Pollution in Victorian and Edwardian Manchester*, (New York and London: Routledge, 2008); J. Whorton, *The Arsenic Century: How Victorian Britain was Poisoned at Home, Work and Play* (Oxford: Oxford University Press, 2010).
8. Bartrip, *The Home Office and the Dangerous Trades*; Whorton, *The Arsenic Century*.
9. J. H. Young, *Pure Food: Securing the Federal Food and Drugs Act of 1906* (Princeton, NJ: Princeton University Press, 1986); D. Carpenter, *The Forging of Bureaucratic Autonomy: Reputations, Networks, and Policy Innovation in Executive Agencies, 1862–1928* (Princeton, NJ: Princeton University Press, 2001); S. Buzzi, P-A Rosental and J-C Devink (eds), *La Santé au Travail 1880--2006* (Paris: La Découverte, 2008); P.-A Dessaux, 'Chemical Expertise and Food Market Regulation in Belle-Epoque France', *History and Technology*, 23:4 (2007), pp. 351–68.
10. J. Whorton, *Before Silent Spring: Pesticides and Public Health in Pre-Ddt America* (Princeton, NJ: Princeton University Press, 1974); F. Davies, 'Pesticides and Toxicology: Episodes in the Evolution of Environmental Risk Assessment (1937–1997)', (PhD dissertation, Yale University, 2001).
11. P. Daniels, *Toxic Drift: Pesticides and Health in the Post-World War II* (New Orleans, LA: Lousiana State University Press, 2007); B. Brooks, *Before Earth Day: The Origins of*

American Environmental Law, 1945–1970 (Lawrence, KS: University of Kansas Press, 2009).

12. D. Milles and R. Müller, 'Die relative Schädlichkeit industrieller Produktion. Zur Geschichte des Grenzwertkonzepts in der Gewerbehygiene', in G. Winter (ed.), *Grenzwerte. Interdisziplinäre Untersuchungen zu einer Rechtsfigur des Umwelt-, Arbeits- und Lebensmittelschutzes* (Düsseldorf: Werner-Verlag, 1986), pp. 227–62; J. Büschenfeld, *Flüsse und Kloaken. Umweltfragen im Zeitalter der Industrialisierung (1870–1914)* (Stuttgart: Klett-Cotta, 1997); C. Reinhardt, 'Boundary Values', in V. Balz, A. v. Schwerin, H. Stoff and B. Wahrig (eds), *Precarious Matters/Prekäre Stoffe, The History of Dangerous and Endangered Substances in the 19th and 20th Centuries* (Berlin: Pre-print series of the Max Planck Institute for the History of Science No. 356, 2008), pp. 39–50; B. Bächi and C. Reinhardt (eds), 'Regulierungswissen', Special Issue of *Berichte zur Wissenschaftsgeschichte*, 33:4 (2010), pp. 345–454.
13. Whorton, *Before Silent Spring*.
14. C. Sellers, *Hazards of the Job: From Industrial Disease to Environmental Science* (Chapel Hill, NC: The University of North Carolina Press, 1997).
15. S. Hays, *Beauty, Health, and Permanence: Environmental Politics in the United States, 1955–1985* (Cambridge: Cambridge University Press, 1989); G. Colten and P. Skinner, *The Road to Love Canal: Managing Industrial Waste before EPA* (Austin, TX: University of Texas Press, 1995); G. E. Markowitz and D. Rosner, *Deceit and Denial: The Deadly Politics of Industrial Pollution* (Berkeley, CA: University of California Press, 2002); Davies, 'Pesticides and Toxicology'; D. Horowitz, *The Anxieties of Affluence. Critiques of American Consumer Culture, 1939–1979* (Amherst and Boston, MA: University of Massachusettes Press, 2005).
16. C. Bosso, *Pesticides and Politics: The Life Cycle of a Public Issue* (Pittsburgh, PA: University of Pittsburgh Press, 1990). Cranor, *Regulating Toxic Substances*; J. Szarka, *The Shaping of Environmental Policy in France* (Oxford: Berghahn Books, 2002).
17. *IARC Monographs on the Evaluation of Carcinogenic Risk of Chemicals to Man,* Vol 1, (Geneva: IARC/WHO, 1972); C. Agathe and J. D. Wilbour, 'IARC Programme on the Evaluation of Carcinogenic Risk of Chemicals to Man', in P. Bucalossi, U. Veronesi and N. Cascinelli (eds), *Proceedings of the Eleventh International Cancer Congress* (Amsterdam: Excerpta Medica, 1975), pp. 108–13; WHO, *Report of a WHO Scientific Group on the Assessment of the Carcinogenecity and Mutagenecity of Chemicals*, WHO technical report (Geneva: WHO, 1974).
18. Heyvaert, 'Reconceptualizing Risk Assessment'; H. Rothstein, A. Irwin, S. Yearley and E. Mc Carthy, 'Regulatory Science, Europeanization, and the Control of Agrochemicals, Science', *Technology and Human Values*, 24:2 (1999), pp. 241–64.
19. Davies, 'Pesticides and Toxicology'.
20. S. Frickel, *Chemical Consequences: Environmental Mutagens, Scientist Activism, and the Rise of Genetic Toxicology* (New Brunswick, NJ: Rutgers University Press, 2004).
21. S. Boudia in this book, p. 71–88.
22. National Research Council, *Risk Assessment in the Federal Government: Managing the Process* (Washington, DC: National Academy Press, 1983).
23. D. Demortain, *Scientists and the Regulation of Risk: Standardising Control* (Cheltenham UK and Northampton US: Edward Elgar Publishers, 2011).
24. N. Pellow, *Resisting Global Toxics: Transnational Movements for Environmental Justice*, (Cambridge, MA and London: The MIT Press, 2007).

25. P. Brown and E. Mikkelsen, *No Safe Place. Toxic Waste, Leukemia and Community Action* (Berkley, CA: University of California Press, 1990); P. Brown, *Toxic Exposures. Contested Illnesses and the Environmental Health Movement* (New York: Columbia University Press, 2007).
26. Heyvaert, 'Reconceptualizing Risk Assessment', p. 139.
27. Applegate, 'Synthesizing TSACA and REACH'; R. Denison, *Ten Essential Elements in TSCA Reform* (Washington, DC: Environmental Law Institute, 2009 at http://www.edf.org/documents/9279_Denison_10_Elements_TSCA_Reform.pdf [accessed 24 September 2012].
28. United Nations Industrial Development Organization.
29. Selin, 'Coalition Politics and Chemicals Management'.
30. N. Sachs, 'Jumping the Pond: Transnational Law and the Future of Chemical Regulation', *Vanderbilt Law Review*, 62:6 (2009), pp. 1817–69; D. Scoot, 'Testing Toxicity: Proof and Precaution Canada's Chemicals Management Plan', *Recueil*, 18:1 (2009), pp. 59–76.
31. Applegate, 'Synthesizing TSACA and REACH'; N. Eckley and H. Selin, 'All Talk, Little Action: Precaution and European Chemicals Regulation', *Journal of European Public Policy*, 11:1 (2004), pp. 78–105.
32. Scoot, 'Testing Toxicity'; P. Goldman and J. Wagner, 'Trading away Public Health: WTO Obstacles to Effective Toxics Control', *Journal of Public Health Policy*, 21:3 (2000), pp. 260–7.
33. M. Foucault, *The Birth of Biopolitics. Lectures at the Collège de France, 1978–1979* (Basingstoke: Palgrave Macmillan, 2008).
34. M. Dean, *Governmentality. Power and Rule in Modern Society* (London: Sage, 1999).
35. N. Rose, *Powers of Freedom: Reframing Political Thought* (Cambridge: Cambridge University Press, 1999); M. Dean, *Governmentality*.
36. D. Chandrasekhar, 'Understanding Stakeholder Participation in Post-Disaster Recovery (Case Study: Nagapattinam, India)' (PhD Dissertation, University of Illinois, 2010).
37. S. Revet, 'De la vulnérabilité aux vulnérables. Approche critique d'une notion performative', in S. Becerra and A. Peletier (eds), *Risques et environnement: recherches interdisciplinaires sur la vulnérabilité des sociétés*, (Paris: L'Harmattan, 2009), pp. 89–99.
38. See United Nations Office for Disaster Risk Reduction: http://www.unisdr.org/who-we-are/unisdr-in-un [accessed 24 September 2012].
39. R. Carson, *Silent Spring* (Boston, MA: Houghton Mifflin, 1962), p. 6.
40. N. Langston, *Toxic Bodies: Hormone Disruptors and the Legacy of DES* (New Haven, CT: Yale University Press, 2009).

1 Sellers, 'The Cold War over the Worker's Body: Cross-National Clashes over Maximum Allowable Concentrations in the Post-World War II Era'

1. International Union of Pure and Applied Chemistry, Applied Chemistry Section & Permanent Commission and International Association on Industrial Medicine, *Proceedings of the International Symposium on Maximum Allowable Concentrations of Toxic Substances in Industry, Held in Prague, Czechoslovakia, April 1959* (London: Butterworths, 1961).

2. S. Boehmer-Christiansen, 'The Precautionary Principle in Germany – Enabling Government', in T. O'Riordan and J. Cameron (eds), *Interpreting the Precautionary Principle* (London: Earthscan, 1994), pp. 31–60; D. Rosner and G. Markowitz, 'Industry Challenges to the Principles of Prevention in Public Health: the Precautionary Principle in Historical Perspective', *Public Health Reports*, 117 (2002), pp. 501–12; F. Uekotter, 'Der Wert des Vorsorgeprinzips: Die Gescheiterte Bekampfung der Automobilasbases in Deutschland und den USA vor 1945', *Technikgeschichte*, 69:1 (2002), pp. 41–72; N. Langston, 'The Retreat from Precaution: Regulating Diethylstilbestrol (DES), Endocrine Disruptors, and Environmental Health', *Environmental History*, 13:1 (2008), pp. 41–65; C. Omnès, 'De la perception du risque professionnel aux pratiques de prévention : la construction d'un risque acceptable', *Revue d'histoire moderne et contemporaine*, 56:1 (2009), pp. 61–82.
3. C. Sellers, *Hazards of the Job*; C. Warren, *Brush with Death: A Social History of Lead Poisoning* (Baltimore, MD: Johns Hopkins University Press, 2002); G. E. Markowitz and D. Rosner, *Deceit and Denial.*
4. J. Rainhorn, 'The Fight Against Lead Poisoning in the Paint Industry: A Comparison of the French and American Experiences 1900 – 1940', in *Proceedings of 4th International Conference on History of Occupational and Environmental Health* (San Francisco, CA: University of California Press and the California Digital Library, 2012); J. Melling and C. Sellers (eds), *Dangerous Trade: Histories of Industrial Hazard across a Globalizing World* (Philadelphia, PA: Temple University Press, 2011).
5. D. T. Rodgers, *Atlantic Crossings: Social Politics in a Progressive Age* (Cambridge, MA and London: Belknap Press of Harvard University, 2000); A. Grieco (ed.), *Origins of Occupational Health Associations in the World* (Amsterdam: Elsevier, 2003); P. Rosental, 'Health and Safety at Work: An Issue in Transnational History Introduction', *Journal of Modern European History,* 7:2 (2009); pp. 169–73; P. Rosental, T. Cayet and M. Thebaud-Sorger, 'How International Organizations Compete Occupational Safety and Health at the ILO: a Diplomacy of Expertise', *Journal of Modern European History*, 7:2 (2009), pp. 174–96; Melling and Sellers, *Dangerous Trade.*
6. C. Hayden, 'Bioprospecting's Representational Dilemma', *Science as Culture*, 14:2 (2005), pp. 185–200.
7. Sellers, *Hazards of the Job.*
8. Hayden, 'Bioprospecting's Representational Dilemma'.
9. B. Adam, *Timescapes of Modernity* (New York: Routledge, 1998); B. Adam, S. Allan and C. Carter, *Environmental Risks and the Media* (New York: Routledge, 2000); J. Doyle, 'Picturing the Clima(c)tic: Greenpeace and the Representational Politics of Climate Change Communication', *Science as Culture*, 16:2 (2007), pp. 129–50.
10. S. N. Tesh, *Uncertain Hazards* (Ithaca: Cornell University Press, 2000); Murphy, *Sick Building Syndrome and the Problem of Uncertainty.*
11. A. Birn, 'The Stages of International (Global) Health: Histories of Success or Successes of History?', *Global Public Health*, 4:1 (2009), pp. 50–68.
12. ICOH, *Proceedings of the XII International Congress on Occupational Health: Helsinki, 1.–6. 7. 1957* (Helsinki: Valtioneuvoston kirjapaino, 1958).
13. A. Hamilton, *Exploring the Dangerous Trades; the Autobiography of Alice Hamilton, M. D.* (Boston, MA: Little, Brown and Company, 1943).
14. T. M. Legge, *Shaw Lectures on Thirty Years Experience of Industrial Maladies. Delivered Before the Royal Society of Arts, February and March, 1929* (London: The Society, 1929); L. Teleky, *History of Factory and Mine Hygiene* (New York: Columbia Univ. Press, 1948).

Rosental, Cayet and Thebaud-Sorger, 'How International Organizations Compete Occupational Safety and Health at the ILO'.

15. M. Robert and L. Parmeggiani, *Fifty Years of International Collaboration in Occupational Safety and Health* (Geneva: International Occupational Safety and Health Information Centre, 1969); J. Heitmann, 'The ILO and the Regulation of White Lead in Britain during the Interwar Years: An Examination of International and National Campaigns in Occupational Health', *Labour History Review*, 69:3 (2004), pp. 267–84; Rainhorn, 'The Fight Against Lead Poisoning in the Paint Industry'.
16. Rosental, 'Health and Safety at Work'.
17. E. T. Beintker, *The Medical Inspection of Labour. Report of the Meeting of Medical Inspectors of Medical Inspectors of Labour, Held in Düsseldorf, 15–16 September 1926* (Geneva: International Labour Office, 1926); Teleky, *History of Factory and Mine Hygiene*; B. Harrison, *Not Only the 'Dangerous Trades': Women's Work and Health in Britain, 1880–1914* (London: Taylor & Francis, 1996).
18. A. Uytdenhoef and R. Genot, 'Pollution sur les lieux de travail', *Acta Tuberculosea Et Pneumologica Belgica*, 58:5 (1967), pp. 443–58; H. Bastenier, H. *La Formation de médecins de travail en Belgique*, *Bruxelles Médical*, 50:4 (1970), pp. 273–6; P. Hublet, 'L'institution des services médicaux de travail', *Bruxelles Médical*, 50:4 (1970), pp. 239–42.
19. R. Barthe, 'Histoire de la médecine du travail française de 1915 à 1940, I, Chapitre II. Services médicaux pilotes', *Concours Médical*, 77:17 (1955), pp. 1753–8; R. Barthe, 'Histoire de la médecine du travail française', *Concours Médical*, 77:21 (1955), pp. 2177–86; C. Omnès and L. Pitti (eds), *Cultures du risque au travail et pratiques de prévention au XXe siècle: La France au regard des pays voisins* (Rennes: Presses Universitaires de Rennes, 1995).
20. P. Boulange, 'L'activité d'une commission de hygiène et sécurité au cours de 1953 dans une usine sidérurgique', *Revue Médicale de Nancy*, 79 (1954), pp. 865–72; C. Adam and L. Kusterer, 'Le fonctionnement des services médicaux interentreprises 10 ans après promulgation de la loi du 11 oct, 1946', *Archives des maladies professionnelles de médecine du travail et de sécurité sociale*, 17 (1956), pp. 180–91.
21. D. Milles, 'Occupational Illnesses to Be Compensated, or Workers' Diseases to Be Eradicated?, *Dynamis*, 17 (1987), pp. 275–305; R. N. Proctor, *The Nazi War on Cancer* (Princeton, NJ: Princeton University Press, 2000); M. Lengwiler, 'Internationale Expertennetzwerke und nationale Sozialstaatsgeschichte: Versicherung der Silikose in Deutschland und der Schweiz (1900–1945)', *Journal of Modern European History*, 7 (2009), pp. 197–218.
22. Robert and Parmeggiani, *Fifty Years of International Collaboration.*
23. Sellers, *Hazards of the Job.*
24. T. M. Porter, *Trust in Numbers* (Princeton, NJ: Princeton University Press, 1996).
25. C. Sellers, 'The Public Health Service's Office of Industrial Hygiene and the Transformation of Industrial Medicine', *Bulletin of the History of Medicine*, 65:1 (1991), pp. 42–73.
26. B. I. Castleman and G. E. Ziem, 'Corporate Influence on Threshold Limit Values', *American Journal of Industrial Medicine*, 13:5, pp. 531–59; S. Czerczak, J. A. Indulski and Z. Kowalski, 'Occupational Exposure Standards. Historical Outline and Present State', *Polish Journal of Occupational Medicine and Environmental Health*, 6:1 (1993), pp. 1–18; J. C. Ward and C. Warren, *Silent Victories: The History and Practice of Public Health in Twentieth-Century America* (Oxford and New York: Oxford University Press, 2007).

27. I. Izmerov, 'Development of Occupational Health in Russia – Historical Background', in A. Greico (ed.), *Origins of Occupational Health Associations in the World* (Amsterdam: Elsevier, 2003), pp. 173–82; D. S. Sharp and M. B. Tauger, "Natan 'Nikolai" Abramovich Vigdorchik (1874–1954): Social Activism and Public Health in early 20th-Century Russia', *Journal of Medical Biography*, 17:2 (2009), pp. 75–80.
28. A. Hamilton, 'Industrial Hygiene in Moscow', *Journal of Industrial Hygiene*, 6:3 (1925), pp. 47–61.
29. Sharp and Tauger, 'Natan "Nikolai" Abramovich Vigdorchik'.
30. L. Carozzi and International Labor Office – Industrial Health Service, *Industrial Medicine in the U. S. S. R.* (Geneva: International Labor Office, 1936); H. J. Magnuson et al., 'Industrial Toxicology in the Soviet Union – Theoretical and Applied', *American Industrial Hygiene Association Journal*, 25 (1964), pp. 185–97; L. Yanin. (ed.), *Labor Hygiene and Industrial Sanitation; a Collection of the Principal Official Publications* (Jerusalem: Israel Program for Scientific Translations, 1966).
31. United States Department of Labor. Bureau of Labor Statistics, *Occupational Health Services in the Soviet Union: A Description and Appraisal* (Washington, DC: Government Printing Office, 1959); A. Dementyeva, A., *Labour Protection in the USSR* (Moscow: Trade-Union Publishing House, 1969).
32. N. Vigdortchik, 'Lead Intoxication in the Etiology of Hypertonia', *Journal of Industrial Hygiene*, 17 (1935), pp. 1–6.
33. R. Truhaut, *Food Additive Control in France* (Rome: Food and Agriculture Organization of the United Nations, 1963); F. R. Davis, 'Pesticides and Toxicology'.
34. S. Smeljanskij, 'Hygienische Begrundung der Normativvorschriften bezuglich maximalzulassiger Konzentrationen Toxischer Stoffe in der Luft der Betriebesraume', in *Reports*, International Congress on Occupational Health (Helsinki: ICOH, 1957), pp. 80–83; G. Windholz, 'Pavlov and the Pavlovians in the Laboratory', *Journal of the History of the Behavioral Sciences*, 26 (1990), pp. 64–74; D. P. Todes, *Pavlov's Physiology Factory: Experiment, Interpretation, Laboratory Enterprise* (Baltimore, MD: Johns Hopkins University Press, 2002).
35. S. Smeljanskij, 'Grundsatze und Praxis der Anordnung uber Maximal Zulassigen Konzentrationen Toxischer Stoffe in der Luft der Industriellen Beriebsraume in der U.S.S.R', in *Proceedings of International Symposium on Maximum Allowable Concentrations of Toxic Substances in Industry* (Prague: Butterworths, 1961), pp. 117–20.
36. C. Cavailles, 'Les moyens matériels mis à la disposition du médecin du Travail. Appareillages et locaux', *Archives des maladies professionnelles de médecine du travail et de sécurité sociale*, 19:2 (1957), pp. 188–9.
37. R. Truhaut, *Contribution à l'étude des facteurs chimiques de cancérisation. Le problème des substances cancérigènes endogènes* (Paris: Société d'édition d'enseignement supérieur, 1947); R. Truhaut, *Principales données actuelles sur les facteurs chimiques de cancérisation* (Paris: Presses documentaires, 1953); R. Truhaut, 'Maximum Allowable Concentrations, Second International Symposium, 1963', *Archives of Environmental Health*, 8, (1964), pp. 487–91; R. Truhaut, *Notice sur les titres et travaux scientifiques de René Truhaut* (Paris: Société d'édition d'enseignement supérieur, 1967).
38. ICOH, *Proceedings of the XII International Congress on Occupational Health: Helsinki, 1.–6. 7. 1957.*
39. R. Truhaut, 'Allocution de l'ouverture', in *Proceedings of International Symposium on Maximum Allowable Concentrations of Toxic Substances in Industry*, (Prague: Butterworths, 1961), pp. 17–20.

40. J. Barnes, 'Experimental Methods Used as a Basis for Determining Maximum Allowable Concentrations', in *Proceeding, International Symposium on Maximum Allowable Concentrations of Toxic Substances in Industry* (Prague: Butterworths, 1961), pp 97–107; W. von Oettingen, 'Methods Used by the Committee Z37 of the American Standards Association in Establishing Maximal Acceptable Concentrations of Toxic Dusts and Gases', in *Proceedings, International Symposium on Maximum Allowable Concentrations of Toxic Substances in Industry* (Prague: Butterworths, 1961), pp. 115–6.
41. H. E. Stokinger, 'Advances in Industrial Toxicology for the Year 1955', *A.M.A. Archives of Industrial Health*, 14:2 (1956), pp. 206–12; H. E. Stokinger, 'Methods of Establishing Threshold Limits; Prepared Discussion', *American Industrial Hygiene Association Quarterly*, 17:3 (1956), pp. 284–6.
42. F. Kaloyanova-Semeonova, 'The Effect of Individual Sensitivity and Physical Exertion in Poisoning with Small Doses of Toxic Substances', in *Proceedings, International Symposium on Maximum Allowable Concentrations of Toxic Substances in Industry* (Prague: Butterworths, 1961), pp. 83–4; A. A. Letavet, 'Basic Concepts for Maximum Allowable Concentrations of Toxic Substances', in *Proceedings, International Symposium on Maximum Allowable Concentrations of Toxic Substances in Industry*, p. 000–000; E. Lyublina, 'Some Methods Used in Establishing the Maximum Allowable Concentrations', in *Proceedings of International Symposium on Maximum Allowable Concentrations of Toxic Substances in Industry*, pp. 109–14.
43. H. E. Stokinger, 'Standards for Safeguarding the Health of the Industrial Worker', *Public Health Reports*, 70:1 (January 1955), pp. 1–11, on p. 1.
44. A. Uytdenhoef, 'Remarques de l'administration belge au sujet du problème de la concentration tolérable des matières nocives dans l'atmosphère des lieux de travail', in *Proceedings of International Symposium on Maximum Allowable Concentrations of Toxic Substances in Industry* (London: Butterworths, 1961); Uytdenhoef and Genot, 'Pollution sur les lieux de travail'.
45. G. Burger, 'Aspects of Safety Limits in Industry', in *Proceedings of International Symposium on Maximum Allowable Concentrations of Toxic Substances in Industry*, pp. 91–6.
46. H. Hardy, *Diary* (Cambridge, MA: Harriet Hardy Papers, Schlesinger Library, Radcliffe College, 1959).
47. Magnuson et al., 'Industrial Toxicology in the Soviet Union – Theoretical and Applied'.
48. A. N. Kachur et al., 'Environmental Conditions in the Rudnaya River Watershed – a Compilation of Soviet and post-Soviet Era Sampling Around a Lead Smelter in the Russian Far East', *The Science of the Total Environment*, 303:1–2 (2003), pp. 171–85; D. A. Wilson et al., 'Lung, Liver and Bone Cancer Mortality after Plutonium Exposure in Beagle Dogs and Nuclear Workers', *Health Physics*, 98:1 (2010), pp. 42–52.
49. Truhaut, 'Maximum Allowable Concentrations, Second International Symposium, 1963'; Joint ILO/WHO Committee on Occupational Health, *Permissible Levels of Occupational Exposure to Airborne Toxic Substances* (Geneva: World Health Organization, 1969); International Labour Office, *Encyclopedia of Occupational Health and Safety* (Geneva: ILO, 1971).
50. T. F. Hatch, 'The Role of Permissible Limits for Hazardous Airborne Substances in the Working Environment in the Prevention of Occupational Disease', *Bulletin of the World Health Organization*, 47:7 (1972), pp. 151–9.
51. Hatch, 'The Role of Permissable Limits for Hazardous Airborne Substances in the Working Environment in the Prevention of Occupational Disease'.

2 Jas, 'Adapting to "Reality": The Emergence of an International Expertise on Food Additives and Contaminants in the 1950s and early 1960s'

1. FAO/WHO, *Joint FAO/WHO Expert Meeting to Review Toxicological and Health Aspects of Bisphenol A. Summary Report including Report of Stakeholder on Bisphenol A* (Geneva: World Health Organization, 2010).
2. S. Vogel, 'The Battles over Bisphenol A', *DefendingScience,* 2008, at http://www.defendingscience.org/case_studies/Battles-Over-Bisphenol-A.cfm [accessed 11 July 2012]; D. J. DeNoon, 'Bottles and Can Chemical Linked to Heart Problems, WebMD', 2010, at http://www.webmd.boots.com/heart-disease/news/20100114/bottle-and-can-chemical-linked-to-heart-problems [accessed 11 July 2012].
3. Vogel, 'The Battles over Bisphenol A'; E. Lane, 'In Assessing Benefit and Risks of Controversial Chemicals, Experts Find Little Certainty', American Association for the Advancement of Science, 2010, at http://www.aaas.org/news/releases/2010/1115bpa.shtml?sa_campaign=Internal_Ads/AAAS/RSS_News/2010-11-15 [accessed 11 July 2012]; S. Vogel, 'BPA Safety Question an Opportunity for Reform', The Pumphandle, 2010, at http://thepumphandle.wordpress.com/2010/01/20/bpa-safety-question-an-opportunity-for-reform/ [accessed 11 July 2012].
4. Brickman, Jasanoff and Ilgen, *Controlling Chemicals*; G. Hoberg, 'Risk, Science and Politics: Alachor Regulation in Canada and the United States', *Canadian Journal of Political Science*, 23:2 (1990), pp. 257–77.
5. B. Gillespie, D. Eva and R. Johnston, 'Carcinogenic Risk Assessment in the United States and Great Britain: The Case of Aldrin/Dieldrin', *Social Studies of Science*, 9:3 (1979), pp. 265–301; J. Abraham, 'Scientific Standards and Institutional Interests: Carcinogenic Risk Assessment of Benoxaprofen in the UK and US', *Social Studies of Science*, 23:3 (1993), pp. 387–444; P. van Zwanenberg and E. Millstone, 'Beyond Skeptical Relativism: Evaluating the Social Constructions of Expert Risk Assessments', *Science, Technology and Human Values*, 25:3 (2000), pp. 259–82.
6. J. Abraham and T. Reed, 'Progress, Innovation and Regulatory Science in Drug Development: The Politics of International Standard-Setting', *Social Studies of Science*, 32:3 (2002), pp. 337–69; D. Demortain, 'From drug crises to regulatory change: The mediation of expertise', *Health, Risk and Society*, 10:1 (2008), pp. 37–51; R. Diggle, 'Regulatory Science and Uncertainty in the Risk Assessment of Pesticide Residues' (PhD dissertation, University of Nottingham, 2010).
7. The latter category consists primarily of pesticide residues, but also of residues from other sources such as antibiotics and food supplements given to animals, manufacturing processes and food containers.
8. Young, *Pure Food*; M. French and J. Phillips, *Cheated Not Poisoned? Food Regulation in the United Kingdom 1875–1938* (Manchester: Manchester University Press, 2009); Dessaux, 'Chemical Expertise and Food Market Regulation in Belle-Epoque France'.
9. P. Zylberman, 'Making Food Safety an Issue: Internationalized Food Politics and French Public Health from the 1970s to the Present', *Medical History*, 48:1 (2004), pp. 1–28; N. Jas, 'Santé publique et marché des agrumes dans la CEE des années 1960', in L. Bonnaud and N. Joly, *L'alimentation sous contrôle. Tracer, Auditer, Conseiller* (Dijon: Educagri/Quae, 2012), pp. 94–108. For later development of international regimes of regulation

of food additives and contaminants see Millstone and van Zwanenberg, 'The Evolution of Food Safety Policy-Making Institutions in the UK, EU and Codex Alimentarius'.

10. FAO/WHO, *Joint FAO/WHO Conference on Food Additives, Geneva 19–22 September 1955, Report* (Geneva: World Health Organization, 1956).
11. V. Wunderlich, 'Zur Entstehungsgeschichte der Druckrey-Kuepfmueller-Schriften (1948–9). Dosis und Wirkung bei krebserzeugenden Stoffen', *Medizinhistorisches Journal*, 40 (2005), pp. 369–97.
12. H. Druckrey, 'Cancer Prevention', *Acta-Unio Internationalis Contra Cancrum*, 13 (1957), pp. 14–25, on p. 14.
13. Druckrey, 'Cancer Prevention', p. 14.
14. Druckrey, 'Cancer Prevention', p. 16.
15. Druckrey, 'Cancer Prevention', p. 16.
16. R. Truhaut (ed.), *Pollution de l'air. Comptes-Rendus du Colloque International de Royaumont, Avril 1960, tenu sous l'égide du Comité Eurotox, Comité Européen Permanent de Recherches pour la Protection des Populations contre les Risques d'Intoxication à Long Terme* (Paris : SEDES, 1961), pp. 115–45.
17. 'Panel Discussion on Regulations Governing Dyes and Other Additives in Foods', *Acta-Unio Internationalis Contra Cancrum*, 13 (1957), pp. 145–64.
18. C. Sellers, 'Discovering Environmental Cancer: Wilhelm Hueper, Post-World War II Epidemiology, and the Vanishing Clinician's Eye', *American Journal of Public Health*, 87:11 (1987), pp. 1824–35.
19. R. Truhaut, 'Les Risques de Cancérisation pouvant résulter de la Présence de Substances Etrangères dans les Aliment: Essai d'Etablissement d'un Doctrine', *Acta-Unio Internationalis Contra Cancrum*, 13 (1957), pp. 28–40, on p. 35.
20. Truhaut, 'Les Risques de Cancérisation pouvant résulter de la Présence de Substances Etrangères dans les Aliment', p. 35.
21. R. Truhaut, 'Les Risques de Cancérisation pouvant résulter de la Présence de Substances Etrangères dans les Aliments: Position des Problèmes. Résolutions Adoptées au cours des Réunions Internationales Antérieures', *Acta-Unio Internationalis Contra Cancrum*, 13 (1957), pp. 210–20, on p. 214.
22. Truhaut, *Pollution de l'air*, p. 8.
23. Truhaut, *Pollution de l'air*, p. 153.
24. Truhaut, *Pollution de l'air*.
25. Truhaut, *Pollution de l'air*, pp. 190–209.
26. P. Groenewegen, 'Accommodating Science to External Demands: the Emergence of Dutch Toxicology', *Science, Technology and Human Values*, 27:4 (2002), pp. 479–98.
27. FAO/WHO, *Joint FAO/WHO Conference on Food Additives*, p. 5.
28. FAO/WHO, *Joint FAO/WHO Conference on Food Additives*, p. 6
29. FAO/WHO, *Joint FAO/WHO Conference on Food Additives*, p. 6
30. FAO/WHO, *Joint FAO/WHO Conference on Food Additives*, p. 9.
31. FAO/WHO, *Joint FAO/WHO Conference on Food Additives*, p. 6.
32. FAO/WHO, *Joint FAO/WHO Conference on Food Additives*, p. 6.
33. FAO/WHO, *Evaluation of the Toxicity of a Number of Antimicrobials and Antioxidants. Sixth Report of the Joint FAO/WHO Expert Committee on Food Additives* (Geneva: World Health Organization, 1962), p. 6.
34. FAO/WHO, *Evaluation of the Toxicity of a Number of Antimicrobials*, p. 6.
35. FAO/WHO, *Second Joint FAO/WHO Conference on Food Additives, Rome, 24–25 June 1963, Report* (Geneva: World Health Organization, 1963), p. 8.

36. R. Truhaut, 'The Concept of the Acceptable Daily Intake: an Historical Review', *Food Additives and Contaminants*, 8:2 (1991), pp. 151–62. See also, E. Poulsen, 'René Truhaut and the Acceptable Daily Intake: A Personal Note', *Teratogenesis, Carcinogenesis, and Mutagenesis*, 15:6 (1995), pp. 273–5.
37. Truhaut, 'The Concept of the Acceptable Daily Intake', p. 153.
38. Truhaut, 'The Concept of the Acceptable Daily Intake', p. 153.
39. FAO/WHO, *General Principles Governing the Use of Food Additives. First Report of the Joint FAO/WHO Expert Committee on Food Additives* (Geneva: World Health Organization, 1957), p. 11.
40. FAO/WHO, *Procedures for the Testing of Intentional Food Additives to Establish their Safety for Use. Second Report of the Joint FAO/WHO Expert Committee on Food Additives* (Geneva: World Health Organization, 1958), p. 17.
41. FAO/WHO, *Procedures for the Testing of Intentional Food Additives*, p. 17.
42. Truhaut, 'The Concept of the Acceptable Daily Intake', p. 154.
43. Truhaut, *Pollution de l'air*, p. 126. Truhaut, 'The Concept of the Acceptable Daily Intake', p. 154.
44. FAO/WHO, *Evaluation of the Toxicity of a Number of Antimicrobials*, p. 6.
45. FAO/WHO, *Principles Governing Consumer Safety in Relation To Pesticide Residues. Report of a Meeting of a WHO Expert Committee on Pesticide Residues held jointly with the FAO Panel of Experts on the Use of Pesticides in Agriculture* (Geneva: World Health Organization, 1962), p. 5.
46. FAO/WHO, *Principles Governing Consumer Safety in Relation To Pesticide Residues*, p. 5.
47. FAO/WHO, *Principles Governing Consumer Safety in Relation To Pesticide Residues*, p. 5.
48. FAO/WHO, *Second Joint FAO/WHO Conference on Food Additives*, p. 8.
49. Poulsen, 'René Truhaut', p. 274.
50. FAO/WHO, *Procedures for the Testing of Intentional Food Additives*, p. 5.
51. E. Poulsen, 'Safety Evaluation of Substances Consumed as Technical Ingredients (food additives)', *Food Additives and Contaminants*, 8:2 (1991), pp. 125–34; A. G. Renwick, 'Safety Factors and Establishment of Acceptable Daily Intakes', *Food Additives and Contaminants*, 8:2 (1991), pp. 135–50.
52. FAO/WHO, *Procedures for the Testing of Intentional Food Additives*, p. 8.
53. Truhaut, *Pollution de l'air*, p. 126.
54. 'Symposium on Food Additives', p. 201. Truhaut, *Pollution de l'air*, p. 8.
55. FAO/WHO, *Procedures for the Testing of Intentional Food Additives*, p. 17.
56. FAO/WHO, *Procedures for the Testing of Intentional Food Additives*, p. 17.
57. FAO/WHO, *Evaluation of the Carcinogenic Hazards of Food Additives. Fifth Report of the Joint FAO/WHO Expert Committee on Food Additives* (Geneva: World Health Organization, 1961), p. 4.
58. FAO/WHO, *Evaluation of the Carcinogenic Hazards of Food Additives*, p. 6.
59. FAO/WHO, *Evaluation of the Carcinogenic Hazards of Food Additives*, p. 6.
60. FAO/WHO, *Evaluation of the Carcinogenic Hazards of Food Additives*, p. 6.
61. FAO/WHO, *Evaluation of the Carcinogenic Hazards of Food Additives*, p. 6.
62. FAO/WHO, *Evaluation of the Carcinogenic Hazards of Food Additives*, p. 7.
63. Truhaut, *Pollution de l'air*, p. 124.
64. FAO/WHO, *Procedures for the Testing of Intentional Food Additives*, p. 12.
65. FAO/WHO, *Evaluation of the Carcinogenic Hazards of Food Additives*, p. 5.

66. FAO/WHO, *Evaluation of the Carcinogenic Hazards of Food Additives*, p. 5.
67. FAO/WHO, *Evaluation of the Carcinogenic Hazards of Food Additives*, p. 19.
68. FAO/WHO, *Evaluation of the Carcinogenic Hazards of Food Additives*, p. 5
69. FAO/WHO, *Evaluation of the Carcinogenic Hazards of Food Additives*, p. 20.
70. FAO/WHO, *Evaluation of the Carcinogenic Hazards of Food Additives*, p. 20.
71. FAO/WHO, *Evaluation of the Carcinogenic Hazards of Food Additives*, p. 20.
72. FAO/WHO, *Evaluation of the Carcinogenic Hazards of Food Additives*, p. 23.
73. This movement was fully achieved in the 1970's. See R. Ricardo-Campbell, *Food Safety Regulation. A Study of the Use and Limitations of Cost-Benefit Analysis* (Washington, DC: AEI-Hoover Policy Studies, 1974).
74. FAO/WHO, *Procedures for the Testing of Intentional Food Additives*, p. 5.
75. FAO/WHO, *Procedures for the Testing of Intentional Food Additives*, p. 18.
76. FAO/WHO, *Evaluation of the Carcinogenic Hazards of Food Additives*, p. 20.
77. FAO/WHO, *Evaluation of the Carcinogenic Hazards of Food Additives*, p. 20.
78. FAO/WHO, *Evaluation of the Carcinogenic Hazards of Food Additives*, p. 20.
79. S. Boudia in this volume, pp. 71–88.
80. Langston, 'The Retreat from Precaution'.
81. WHO Expert Advisory Panel on Food Additives and Contaminants, WHO archives, Geneva, F6/136/3.
82. R. Truhaut, 'Evaluation of Carcinogenic Activity of Food Additives and Contaminants', *Archives of Environmental Health*, 7:4 (1963) pp. 351–8.

3 Boudia, 'From Threshold to Risk: Exposure to Low Doses of Radiation and its Effects on Toxicants Regulation'

1. D. Normile, 'Newsmaker interview: Fukushima Begins 30-Year Odyssey in Radiation Health', *Science*, 333:6043 (2011), pp. 684–5.
2. D. Normile, 'Fukushima Revives the Low-Dose Debate', *Science*, 332:6032 (2011), pp. 908–10, on p. 908.
3. An epidemiological cohort, the Life Span Study (LSS), which included 94,000 atomic-bomb survivors and 27,000 unexposed individual subjects, was selected from residents of Hiroshima and Nagasaki and was followed by the Atomic Bomb Casualty Commission (ABCC), established in 1947 by the US National Academy of Sciences (NAS) with funding from the US Atomic Energy Commission. In 1975, the ABCC was replaced by a Japan–US organization, the Radiation Effects Research Foundation (RERF). This foundation continues to investigate the long-term effects of A-bomb radiation on causes of death and incidences of cancer. On the history of the ABCC, see J. Beatty, 'Genetics in the Atomic Age: the Atomic Bomb Casualty Commission, 1947–1956' in B. R. Keith, J. Maienschein and R. Rainger (eds), *The Expansion of American Biology* (New Brunswick, NJ: Rutgers University Press, 1991), pp. 284–324; M. S. Lindee, *Suffering Made Real: American Science and the Survivors at Hiroshima* (Chicago, IL: University of Chicago Press, 1994). The UNSCEAR 2000 Report assessed data on cancer incidence and mortality up to 1990 among the Life Span Study (LSS) cohort of survivors of the atomic bombings in Japan. See: D. A. Pierce, Y. Shimizu, D. L. Preston et al., 'Studies of the Mortality of Atomic Bomb Survivors. Report 12, Part I. Cancer: 1950–1990', *Radiation Research*, 146:1, (1996), pp.1–27; D. L. Preston, S. Kusumi, M. Tomonaga et al., 'Cancer Incidence in Atomic Bomb Survivors. Part III: Leukemia, Lymphoma and Multiple

Myeloma, 1950–1987', *Radiation Research*, 137:2 (1994), pp. S68–97; D. E. Thompson, K. Mabuchi, E. Ron et al., 'Cancer Incidence in Atomic Bomb Survivors. Part II. Solid Tumors, 1958–1987', *Radiation Research*, 137:2 (1994), pp. S17–67.

4. P. Boyer, *By the Bomb's Early Light: American Thought and Culture at the Dawn of the Atomic Age* (Chapel Hill, NC: The University of North Carolina Press, 1994); B. Balogh, *Chain Reaction: Expert Debate and Public Participation in American Commercial Nuclear Power, 1945–1975* (Cambridge: Cambridge University Press, 1991); B. C. Hacker, *Elements of Controversy: the Atomic Energy Commission and Radiation Safety in Nuclear Weapons Testing, 1947–1974* (Berkeley, CA: University of California Press, 1994); L. S. Wittner, *The Struggle against the Bomb. Volume 1. One World or None: A History of the World Nuclear Disarmament Movement through 1953* (Stanford, CA: Stanford University Press, 1993); L. S. Wittner, *The Struggle against the Bomb. Volume 2. Resisting the Bomb: A History of the World Nuclear Disarmament Movement, 1954–1970* (Stanford, CA: Stanford University Press, 1997); M. Pinault, *Frédéric Joliot-Curie* (Paris, Odile Jacob, 2000); J. Schubert and R. E Lapp, *Radiation: What It Is and How It Affects You* (New York: Viking Press, 1957); R. E. Lapp, *The Voyage of Lucky Dragon* (New York: Harper and Brothers, 1958).
5. C. Kopp, 'The Origins of the American Scientific Debate over Fallout Hazards', *Social Studies of Science*, 9:4 (1979), pp. 403–22; S. Boudia, 'Naissance, extinction et rebonds d'une controverse scientifique: les dangers de la radioactivité pendant la guerre froide', *Mil neuf cent. Revue d'histoire intellectuelle*, 25 (2007), pp. 157–70.
6. E. A. Carlson, 'Genes, Radiation, and Society: The Life and Work of H. J. Muller' (PhD dissertation, Cornell University Press, 1982); A. v. Schwerin, 'From the Atomic Age to Anti Ageing: Shaping the Surface Between the Environment and the Organism', in M. Grote and M. Stadler (eds), *Membranes Surfaces Boundaries. Interstices in the History of Science, Technology and Culture*, (Berlin: Pre-print series of the Max Planck Institute for the History of Science, 2011), pp. 135–57; S. de Chadarvian, 'Mice and the Reactor: The "Genetics Experiment" in 1950s Britain', *Journal of the History of Biology*, 39 (2006), pp. 707–35. See also R. E. Kohler, *Lords of the Fly:* Drosophila *Genetics and the Experimental Life* (Chicago, IL: University of Chicago Press, 1994).
7. The journal *Science* published a series of articles on this matter: H. L. Andrews, 'Radioactive Fallout From Bomb Clouds', *Science*, 122 (1955), pp. 453–6; I. L. Ophel, 'Fallout and the Strontium-90 Hazard', *Science*, New Series, 125:3244 (1957), pp. 381–412.
8. S. Boudia, 'Global Regulation: Controlling and Accepting Radioactivity Risks', History and Technology, 23:4 (2007), pp. 389–406.
9. A. Lacassagne, 'Exposé introductif' in *L'énergie atomique dans ses répercussions sur la vie et la santé* (Paris: L'expansion Editeur, 1955), pp. 1–5, on pp. 2–3.
10. On the history of radiation protection see: L. Taylor, *Organization for Radiation Protection: The Operation of ICPR and NCRP, 1928–1974* (Washington, DC: US Department of Energy, 1979); D. P. Serwer, 'The Rise of Radiation Protection: Science, Medicine and Technology in Society, 1896–1935' (PhD Dissertation, Princeton University, 1977); S. J. Walker, *Permissible Dose: A History of Radiation Protection in the Twentieth Century* (Berkeley, CA: University of California Press, 2000); Boudia, 'Global Regulation: Controlling and Accepting Radioactivity Risks'.
11. G. W. C. Kaye, 'International Recommendation for X-Rays and Radium Protection', *British Journal of Radiology*, 1 (1928), pp. 358–65; G. Failla, 'Radiation Protection', *Radiology*, 19 (1932), pp. 12–21; ICRP, 'International Recommendations for X-Ray and Radium Protection: Stockholm 1929', *Annals of the ICRP/ICRP*, 1 (1959), pp.

62–73; ICRP, 'International Recommendations for X-Ray and Radium Protection: Revised by the International X-Ray and Radium Protection Commission at the Fourth International Congress of Radiology, Zürich, July 1934', *Annals of the ICRP/ICRP*, 1 (1959), pp. 1–5; R. Sievert, 'The Tolerance Dose and the Prevention of Injuries Caused by Ionizing Radiations', *British Journal of Radiology*, 20 (1947), pp. 306–18; R. S. Stone, 'The Concept of a Maximum Permissible Exposure', *Radiology*, 58:5 (1952), pp. 639–61

12. G. Failla, 'US Preparatory Report (9–49). Preliminary Draft of Report of Subcommittee On Permissible Dose from External Radiation of National Committee on Radiation, September 1949', in L. Taylor, *Tripartite Conferences on Radiation Protection: Canada, United Kingdom, United States (1949–1953)* (Washington, DC: National Bureau of Standards, 1984), pp. 2.1–2.7, on pp. 2–7. Underlined in the original text.
13. Georges Beadle was a member of several learned societies, and particularly of the *National Academy of Sciences* (NAS). During the study on the effects of radiation carried out by the NAS, he was chairman of the sub-committee on the genetic effects of atomic radiation.
14. BEAR, *Biological Effects of Atomic Radiation, Report to Public* (Washington, DC: NAS-NRC, 1956); J. D. Hamblin, 'A Dispassionate and Objective Effort: Negotiating the First Study on the Biological Effects of Atomic Radiation', *Journal of the History of Biology*, 40:1 (2007), p. 147–77; Boudia, 'Global Regulation: Controlling and Accepting Radioactivity Risks'.
15. G. Beadle, 'Possible Direct Effects on Man of Low Level Exposures to Ionizing Radiation', one-page memorandum, 1955, cited by J. Caron, 'Edward Lewis and Radioactive Fallout: The Impact of Caltech Biologists on the Debate over Nuclear Weapons Testing in the 1950s and 60s' (PhD Dissertation, California Institute of Technology, 2003).
16. Edward Lewis was laureate of the Nobel Prize in Medicine in 1995.
17. E. Lewis, 'Leukemia and Ionizing Radiation', *Science*, 125:3255 (1957), pp. 965–72.
18. On the development of research on animals in radiobiology, see: K. A. Rader, *Making Mice: Standardizing Animals for American Biomedical Research, 1900–1955* (Princeton, NJ: Princeton University Press, 2004).
19. On the debates which led to the banning of atmospheric tests, see: R. A. Divine, *Blowing on the Wind: The Nuclear Test Ban Debate, 1954–1960* (New York: Oxford University Press, 1981); B. P. Greene, *Eisenhower, Science Advice, and the Nuclear Test-Ban Debate, 1945–1963* (Stanford, CA: Stanford University Press, 2007).
20. Boudia, 'Global Regulation: Controlling and Accepting Radioactivity Risks'.
21. ICRP, 'Recommendations of the International Commission on Radiological Protection (Adopted September 17, 1965)', *Annals of the ICRP/ICRP*, Publication 9 (1965), pp. 1–27.
22. Louis Bugnard, Minutes of the meeting of 2 July 1958 of International Commission on Radiological Protection, Antoine Lacassage's papers, Institut Pasteur Archives, p. 7.
23. On the development of nuclear industry see: G. Hecht, *The Radiance of France: Nuclear Power and National Identity after World War II* (Cambridge, MA: MIT Press, 1998); G. Hecht, *Being Nuclear: Africans and the Global Uranium Trade* (Cambridge, MA: MIT Press 2012); P. Josephson, *Red Atom: Russia's Nuclear Power Program from Stalin to Today* (New York: W. H. Freeman and Company, 2000).
24. J. W. Gofman and A. Tamplin, 'Low Dose Radiation and Cancer', *IEEE Transactions On Nuclear Science*, 17:1 (1970), pp. 1–9.
25. Gofman and Tamplin, 'Low Dose Radiation and Cancer'.

26. J. W. Gofman and A. R Tamplin, 'Federal Radiation Council Guidelines for Radiation Exposure of the Population at Large. Protection or Disaster', in *Hearing Before the Joint Committee on Atomic Energy. Congress of United States. First session on Environmental Effects of Producing Electric Power. October 28, 29, 30, 31; Novembre 4, 5, 6 et 7* (Washington, DC: US Government Printing Office. Washington, 1969).
27. Every industry generally causes residue to be dispersed into the environment and is governed by standards which are supposed to guarantee the innocuous nature of these substances.
28. On the history of nuclear opposition movement: J. Jasper, *Nuclear Politics: Energy and the State in the United States, Sweden and France*, (Princeton, NJ: Princeton University Press, 1990); D. Nelkin and M. Pollack, *The Atom Besieged: Antinuclear Movements in France and Germany* (Cambridge, MA: MIT, 1981), S. Topçu, 'L'agir contestataire à l'épreuve de l'atome. Critique et gouvernement de la critique dans l'histoire de l'énergie nucléaire en France (1968–2008)', (PhD dissertation, EHESS-Paris, 2010).
29. R. Holcomb, 'Radiation Risk: A Scientific Problem?', *Science*, 167:3919 (1970), pp. 853–5.
30. J. W. Gofman, 'L'énergie empoisonnée', *Nouvel Observateur*, June–July (1972), pp. 44–8.
31. P. Boffey, 'Radiation Standards: Are the Right People Making the Decisions?', *Science*, 171:3973 (1971), pp. 780–3.
32. Boudia, 'Global Regulation: Controlling and Accepting Radioactivity Risks'.
33. Argentina, Australia, Belgium, Brazil, Canada, Egypt, United States, France, India, Japan, Great Britain, Mexico, Sweden, Czechoslovakia and USSR. Since then, six other countries have been invited to join the committee: Germany, Indonesia, Peru, Poland, Sudan and China.
34. As part of the process of establishing the standards which are still in force today, the ICRP makes recommendations based on works and data collected by UNSCEAR, by various national science and medical academies and by the international association of radioprotection which groups together national societies. On the basis of these recommendations, the IAEA (internationally) and Euratom (Europe) draw up texts of reference which are then converted into national regulations.
35. Taylor, *Organization for Radiation Protection.*
36. F. Chateauraynaud and D. Torny, *Les sombres précurseurs. Une sociologie pragmatique de l'alerte et du risque* (Paris: Editions de l'EHESS, 1999).
37. Gofman, 'L'énergie empoisonnée', p. 48.
38. Boffey, 'Radiation Standards', p. 783.
39. R. Nader, P. Petkas and K. Blackwell (eds), *Whistle Blowing: The Report of the Conference on Professional Responsibility* (New York: Grossman, 1972).
40. BEIR, *The Effects on Populations of Exposure to Low Levels Ionizing Radiation* (Washington, DC: The National Academies Press, 1972).
41. The distinction between *risk assessment* and *risk management* would only be made as from the 1983 report NAS-NRC report: NAS-NRC, *Risk Assessment in the Federal Government: Managing the Process.*
42. NAS-NRC, *Risk Assessment in the Federal Government: Managing the Process.*
43. NAS-NRC, *Risk Assessment in the Federal Government: Managing the Process*, p. 1156
44. OTA (US Congress, Office of Technology Assessment), *Identifying and Regulating Carcinogens* (Washington, DC: US Government Printing Office, 1987).

45. Epstein, *The Politics of Cancer*; Jasanoff, *The Fifth Branch*; S. Jasanoff (ed.), *Learning from Disaster: Risk Management after Bhopal* (Philadelphia, PA: University of Pennsylvania Press, 1994).
46. BEIR, *The Effects on Populations of Exposure to Low Levels Ionizing Radiation* (Washington, DC: The National Academies Press, 1980); BEIR, *The Effects on Populations of Exposure to Low Levels Ionizing Radiation* (Washington, DC: The National Academies Press, 1972); BEIR, *Health Effects of Exposure to Low Levels of Ionizing Radiation* (Washington, DC: The National Academies Press, 1990); BEIR, *Health Risks from Exposure to Low Levels of Ionizing Radiation* (Washington, DC: The National Academies Press, 2006).
47. NAS-NRC, *Risk Assessment in the Federal Government: Managing the Process.*

4 Henry, '"License to Expose"? Occupational Exposure Limits, Scientific Expertise and State in Contemporary France'

1. E. Henry, *Amiante: un scandale improbable. Sociologie d'un problème public* (Rennes: Presses universitaires de Rennes, 2007).
2. International Labour Organization, *Report of the Meeting of Experts on the Safe Use of Asbestos, Geneva, 11–18 December 1973* (Geneva: ILO, 1974).
3. G. E. Ziem and B. I. Castleman, 'Threshold Limit Values: Historical Perspectives and Current Practice', *Journal of Occupational Medicine*, 31:11 (1989), pp. 910–8.
4. Ziem and Castleman, 'Threshold Limit Values: Historical Perspectives and Current Practice'.
5. F. Ewald, *L'État providence* (Paris: Grasset, 1986).
6. J. M. Paull, 'The Origin and Basis of Threshold Limit Values', *American Journal of Industrial Medicine,* 5:3 (1984), pp. 227–38; S. O. Hansson, *Setting the Limit. Occupational Health Standards and the Limits of Science* (New York: Oxford University Press, 1998).
7. Brickman, Jasanoff and Ilgen, *Controlling Chemicals*; L. Schenk, 'Occupational Exposure Limits in Comparative Perspective: Unity and Diversity Within the European Union' in J. Eriksson, M. Gilek and C. Rudén (eds), *Regulating Chemical Risks* (Dordrecht: Springer Netherlands, 2010), pp. 133–50.
8. R. A. Kagan, *Adversarial Legalism: The American Way of Law* (Cambridge, MA: Harvard University Press, 2001).
9. Brickman, Jasanoff and Ilgen, *Controlling Chemicals*, p. 210.
10. Scientific Committee on Occupational Exposure Limits (SCOEL), *Methodology for the Derivation of Occupational Exposure Limits: Key Documentation (version 6)* (Luxembourg: European Commission, 2009).
11. E. Henry, 'Les politiques de santé au travail : une légitimité fragile', in L. Duclos, G. Groux and O. Mériaux (eds), *Les nouvelles dimensions du politique. Relations professionnelles et régulations sociales* (Paris: LGDJ, 2009), pp. 147–61. An English version of this paper, 'The Uncertain Legitimacy of Occupational Health Policies') is available at http://halshs.archives-ouvertes.fr/halshs-00752892 [accessed 24 September 2012].
12. C. Gilbert and E. Henry. 'Defining Social Problems: Tensions between Discreet Compromise and Publicity', *Revue française de sociologie* (English), 53:1 (2012), pp. 31–54.
13. Brickman, Jasanoff and Ilgen, *Controlling Chemicals*.
14. P. Lascoumes and P. Le Galès (eds), *Gouverner par les instruments* (Paris: Presses de Sciences Po, 2004); P. Lascoumes and P. Le Galès, 'Introduction: Understanding Public

Policy through Its Instruments? From the Nature of Instruments to the Sociology of Public Policy Instrumentation', *Governance*, 20:1 (2007), pp. 1–21 and more generally the articles in this special issue of *Governance*.

15. Hansson, *Setting the Limit;* D. Walters, K. Grodzki and S. Walters, *The Role of Occupational Exposure Limits in the Health and Safety Systems of EU Member States* (London: Health and Safety Executive, 2003).
16. B. Hervé-Bazin, 'Les valeurs limites d'exposition professionnelles en France', in J.-L. Fabiani and J. Theys (eds), *La société vulnérable. Evaluer et maîtriser les risques* (Paris: Presses de l'Ecole normale supérieure, 1987), pp. 307–10.
17. G. Le Ricousse, *L'établissement des valeurs limites d'exposition aux substances toxiques dans les atmosphères de travail. Un tournant dans la prévention des maladies professionnelles en France?* (Paris: INRS, 1985).
18. The CSPRP became the COCT (Conseil d'orientation sur les conditions de travail) in 2009.
19. Interview with the chairman of the G2SAT, 19 May 2009.
20. Circular of 19 July 1982 on acceptable levels for concentrations of certain dangerous substances in the atmosphere of workplaces, *Bulletin Officiel des ministères chargés de l'Emploi et du Travail*, TR 82/38, n° 11453.
21. See INRS, *Valeurs admises pour les concentrations de certaines substances dangereuses dans l'atmosphère des lieux de travail*, INRS, ED 669, November 1985 cited in Hervé-Bazin, 'Les valeurs limites d'exposition professionnelles en France'. Today, a synthesis document has replaced this collection: INRS, *Valeurs limites d'exposition professionnelle aux agents chimiques en France* INRS, ED 984, 2007.
22. The 1982 circular was completed by ten other circulars, up to the one of 21 August 1996, the last to date. See Conseil supérieur de la prévention des risques professionnels, *Bilan des conditions de travail en 2007* (Paris: Ministère du Travail, des Relations sociales et de la Solidarité, 2008), especially ch. 7.
23. Henry, 'Les politiques de santé au travail'.
24. Brickman, Jasanoff, and Ilgen, *Controlling Chemicals*; J-N. Jouzel, 'Une cause sans conséquences. Comparaison des trajectoires politiques des éthers de glycol en France et en Californie' (PhD dissertation, Institut d'études politiques de Grenoble, 2006).
25. On this point, regarding the risk issues, see O. Borraz and C. Gilbert, 'Quand l'Etat prend des risques', in O. Borraz and V. Guiraudon (eds), *Politiques publiques. 1, La France dans la gouvernance européenne* (Paris: Presses de Sciences Po, 2008), pp. 337–57.
26. Conseil supérieur de la prévention des risques professionnels, *Bilan des conditions de travail en 2007*, p. 175.
27. Hervé-Bazin, 'Les valeurs limites d'exposition professionnelles en France', p. 308.
28. For an analysis of a similar process in the specific field of compensation for occupational diseases, see M.-O. Déplaude, 'Codifier les maladies professionnelles: les usages conflictuels de l'expertise médicale', *Revue française de science politique*, 53:5 (2003), pp. 707–35.
29. Interview with the head of the G2SAT, 19 May 2009.
30. S. Kelman, *Regulating America, Regulating Sweden: A Comparative Study of Occupational Safety and Health Policy* (Cambridge, MA: MIT Press, 1981); Brickman, Jasanoff and Ilgen, *Controlling Chemicals*; S. O. Hansson, *Setting the Limit*.
31. On the specific characteristics of expertise in the French context, see C. Restier-Melleray, 'Experts et expertise scientifique. Le cas de la France', *Revue française de science politique*, 40:4, (1990), pp. 546–85. M. Pollak, 'La régulation technologique: le difficile mariage entre le droit et la technologie', *Revue française de science politique*, 32:2 (1982), pp. 165–84.

32. Interview with the head of the G2SAT, 19 May 2009.
33. C. Le Galès and A. Oudiz, 'Contribution méthodologique à la détermination de valeurs limites d'exposition professionnelles à l'amiante. Relation exposition-risque et critères économiques', *Revue d'épidémiologie et de santé publique*, 32:2 (1984), pp. 113–21.
34. Brickman, Jasanoff and Ilgen, *Controlling Chemicals*; Jasanoff, *The Fifth Branch.*
35. See Health and Safety Executive, *Workplace Exposure Limits. EH40/2005* (London: HSE Book, 2005).
36. A. Oudiz and J.-P. Moatti, 'Les valeurs limites d'exposition aux substances cancérigènes: science, gestion du risque et décision', in J.-L. Fabiani and J. Theys (eds), *La société vulnérable. Evaluer et maîtriser les risques*; J-P. Moatti, *Economie de la sécurité. De l'évaluation à la prévention des risques technologiques* (Paris: Inserm-La Documentation française, 1989).
37. D. Benamouzig, *La santé au miroir de l'économie: une histoire de l'économie de la santé en France* (Paris: Presses universitaires de France, 2005).
38. Interview with the head of the G2SAT, 19 May 2009.
39. Interview with the administrative head of the expert committees on exposure limits at the DGT, 5 June 2009.
40. E. Henry, 'Quand l'action publique devient nécessaire. Qu'a signifié 'résoudre' la crise de l'amiante?', *Revue française de science politique*, 54:2 (2004) pp. 289–314; Henry, *Amiante: un scandale improbable.*
41. E. Henry, 'The New Dynamics of Knowledge and the Persistence of Power Relations: The Limited Impact of Important Transformations in Occupational Health Expertise', *Revue française de science politique* (English), 61:4 (2011), pp. 25–43.
42. Conseil supérieur de la prévention des risques professionnels, *Bilan des conditions de travail en 2007*, p. 181.
43. In July 2010, the AFSSET became the Agence Nationale de Sécurité Sanitaire Alimentation, Environnement, Travail (ANSES) when it merged with the Agence Française de Sécurité Sanitaire des Aliments (AFSSA).
44. The Comité Permanent Amiante (CPA) (Asbestos Standing Committee) was a pressure group organized and funded by industries in the sector. Members included scientists, trade unionists and representatives of the government administration. In 1995 it appeared as one of the players impeding decision-making on asbestos.
45. Interview with the administrative head of the expert committee on exposure limits at the DGT, 5 June 2009.
46. Scientific Committee on Occupational Exposure Limits (SCOEL), *Methodology.*
47. See minutes of the meeting of the CSPRP (Commission des risques chimiques, biologiques et des ambiances physiques) 10 April 2006, Records of the Direction Générale du Travail.
48. Conseil supérieur de la prévention des risques professionnels, *Bilan des conditions de travail en 2007*, p. 183.
49. Interview with one of the administrative head of the AFSSET, 16 January 2008.
50. Interview with one of the administrative head working on OELs at the AFSSET, 8 July 2008.
51. AFSSET, *Avis relatif à la proposition de valeurs limites d'exposition à des agents chimiques en milieu professionnel. Evaluation des effets sur la santé et des méthodes de mesure des niveaux d'exposition sur le lieu de travail pour les fibres céramiques réfractaires* (AMaisons-Alfort: AFSSET, 2009), pp. 3–4.

52. AFSSET, *Avis relatif à la proposition de valeurs limites d'exposition à des agents chimiques en milieu professionnel*, p. 21.
53. Interview with one of the administrative heads working on OELs at the AFSSET, 8 July 2008.
54. Minutes of the meeting of the CSPRP Commission des risques chimiques, biologiques et des ambiances physiques, 10 April 2006, Records of the Direction Générale du Travail.
55. Henry, 'The New Dynamics of Knowledge and the Persistence of Power Relations'.

5 Murphy, 'Chemical Infrastructures of the St Clair River'

1. EcoJustice, *Exposing Canada's Chemical Valley: An Investigation of Cumulative Air Pollution Emission in the Sarnia, Ontario Area* (Toronto: EcoJustice, 2007).
2. Sarnia Lambton Economic Partnership, *The Sarnia–Lambton Petrochemical and Refining Complex* (Sarnia: Sarnia Lambton Economic Partnership, 2011).
3. S. L. Star, 'The Ethnography of Infrastructure', *American Behavioral Scientist*, 43:3 (1999), pp. 377–91; S. L. Star and K. Ruhleder, 'Steps Towards an Ecology of Infrastructure', *Information Systems Research*, 7:1 (1996), pp. 111–34.
4. On chemicals and distributed ontologies, see M. Murphy, 'Chemical Regimes of Living', *Environmental History Journal*, 13:4 (2009), pp. 698–703; M. Murphy, 'Distributed Reproduction', in M. Casper and P. Currah (eds), *Corpus: An Interdisciplinary Reader on Bodies and Knowledge* (New York: Palgrave Macmillan, 2011), pp. 21–38.
5. D. Haraway, 'Manifesto for Cyborgs: Science, Technology and Socialist Feminism in the 1980s', *Socialist Review*, 80 (1985), pp. 65–108.
6. D. Nixon, *Slow Violence and the Environmentalism of the Poor* (Cambridge, MA, Harvard University Press, 2011).
7. V. Adams, A. Clarke and M. Murphy, 'Anticipation, Technoscience, Life, Affect, Temporality', *Subjectivities*, 28:1 (2009), pp. 246–65.
8. Environment Canada and Ontario Minister of Environment, 'Pollution of the St. Clair River (Sarnia Area)', (Ontario Ministry of Environment, 1985).
9. Maxwell Cohen, Radio Interview. 'As It Happens', Canadian Broadcast Corporation, 19 March 1979. Available at www.cbc.ca/archives [accessed 27 September 2012].
10. S. Painter, D. Myers and J. Letteros, *Characterization of Data and Data Collection Programs for Assessing Pollutants of Concern to Lake Erie* (Lake Erie Lakewide Management Plan Technical Report Series, 2000) at http://www.epa.gov/glnpo/lakeerie/buia/datachar.pdf [accessed 12 July 2012].
11. S. R. de Solla et al., 'Snapping Turtles (*Chelydra Serpentina*) as Bioindicators in Canadian Areas of Concern in the Great Lakes Basin. I. Spatial Distribution of PBDEs, PCBs and Organochlorine Pesticides in Eggs (*Chelydra Serpentina*)', *Environmental Science and Technology*, 41 (2007), pp. 7252–9.
12. See the environmental justice work of Ada Lockeridge and Ron Plain: D. Scott, 'Confronting Chronic Pollution: Environmental Justice for the Aamjiwnaang First Nation', Paper delivered at the University of Toronto, Jan. 2009; the film *The Beloved Community* by Pamela Calvert/Plain Speech (California Newsreel, 2007); D. Scott, '"Gender Benders": Sex and Law in the Constitution of Polluted Bodies', *Feminist Legal Studies*, 17:3 (2009), pp. 241–65; D. Scott, 'Testing Toxicity: Proof and Precaution in Canada's Chemicals Management Plan', *Review of European Community and International Environmental Law*, 18:1 (2009), pp. 56–76.

13. C. Mackenzie, A. Lockridge and M. Keith, 'Declining Sex Ratio in a First Nation Community', *Environmental Health Perspectives*, 113:10 (2005), pp. 1295–8.
14. EcoJustice, *Exposing Canada's Chemical Valley*.
15. K. Fortun, 'Biopolitics and the Informating of Environmentalism', in K. S. Rajan (ed.), *Lively Capital: Biotechnologies, Ethics, and Governance in Global Markets* (Durham, NC: Duke University Press, 2012), pp. 306–28.
16. Ontario Ministry of Environment Website, 'Environmental Approvals', 2010, at www.ene.gov.on.ca/environment [accessed 12 July 2012].
17. Ecojustice, 'Application for Review Under Section 61 of the Environmental Bill of Rights, 1993: Concerning the Need for Regulatory Amendments for Pollution, Hotspots', 29 January 2009.
18. Aamjiwnaang Bucket Brigade, Report on Sample II, 10 April 2008.
19. Canadian Environmental Protection Act, 1999, *Priority Substances List Assessment Report: Carbon Disulfide* (Ottawa: Environment Canada, 2000), p. 2.
20. Canadian Environmental Protection Act, 1999, *Priority Substances List Assessment Report*, p. 2.
21. Agency for Toxic Substances and Disease Registry, *Toxicological Profile for Carbon Disulfide*. August 1996.
22. Y. L. Wang and X. H. Zhao, 'Occupational Health of Working Women in China', *Asian-Pacific Journal of Public Health*, 1:4 (1987), pp. 66–71.
23. S. L. Jain, 'Living in Prognosis: Toward an Elegiac Politics', *Representations*, 98:1 (2007), pp. 77–92.
24. M. Susiarjo et al., Bisphenol A Exposure in Utero Disrupts Early Oogenesis in the Mouse', *PLoS Genetics*, 3:1 (2007), pp. 63–70.
25. S. L. Jain, *Commodity Violence* (Durham, NC: Duke University Press, forthcoming).
26. A. M. Soto et al., 'The E-Screen Assay as a Tool to Identify Estrogens: An Update on Estrogenic Environmental Pollutants', *Environmental Health Perspectives*, 103:7 (1995), pp. 113–22; A. M. Soto et al., 'P-Nonyl Phenol: An Estrogenic Xenobiotic Released from Modified Polystyrene', *Environmental Health Perspectives*, 92 (1991), pp. 167–73.
27. A. Soto and C. Sonnenschein, *The Society of Cells: Cancer and Control of Cell Proliferation* (New York: Springer-Verlag, 1998).
28. M. A. Cooney et al., 'Household Pesticides and the Risk of Wilms Tumor', *Environmental Health Perspectives*, 115:1 (2007), pp. 134–7; C. R. Sharpe et al., 'Parental Exposures to Pesticides and Risk of Wilms Tumor in Brazil', *American Journal of Epidemiology*, 141:3 (1995), pp. 210–7.
29. P. Mocarelli et al., 'Change in Sex Ratio with Exposure to Dioxin', *Lancet*, 348 (1996), pp. 409–10.
30. D. V. Weseloh, C. Perkarik and S. R. De Solla, 'Spatial Patterns and Rankings of Contaminant Concentrations in Herring Gull Eggs from 15 Sites in the Great Lakes and Connecting Channels, 1998–2002', *Environmental Monitor Assessment*, 113 (2003), pp. 265–84; Environmental Canada, 'Great Lakes Fact Sheet: Fish and Wildlife Health Effects in the Canadian Great Lakes Areas of Concern', (2003); S. R. de Solla, C. A. Bishop and R. J. Brooks, 'Sexually Dimorphic Morphology of Hatchling Snapping Turtles (*Chelydra Serpentina*) from Contaminated and Reference Sites in the Great Lakes and St. Lawrence River Basin, North America', *Environmental Toxicology and Chemistry*, 21:5 (2002), pp. 7252–9; R. J. Kavanagh et al., 'Endocrine Disruption and Altered Gonadal Development in White Perch (*Morone Americana*) from the Lower Great Lakes Region', *Environmental Health Perspectives*, 112:8 (2004), pp. 898–902.

31. J. R. Marantette et al., 'Signatures of Contamination in Invasive Round Gobies (Neogobius Melanostomus): A Double Strike for Ecosystem Health?', *Ecotoxicology and Environmental Safety*, 73:7 (2010), pp. 1755–64.
32. J. R. Marantette et al., 'Signatures of Contamination in Invasive Round Gobies'.
33. L. A. Bowley at al., 'Characterization of Vitellogenin Gene Expression in Round Goby (*Neogobius Melanostomus*) Using a Quantitative Polymerase Chain Reaction Assay', *Environmental Toxicology and Chemistry*, 29:12 (2010), pp. 2751–60.
34. T. George, *Minamata: Pollution and the Struggle for Democracy in Postwar Japan* (Cambridge, MA: Harvard University Press, 2002); B. Walker, *Toxic Archipelago: A History of Industrial Disease in Japan* (Seattle, WA: University of Washington Press. 2011).
35. I. Luginaah, K. Smith and A. Lockridge, 'Surrounded by Chemical Valley and "Living in a Bubble": The Case of the Aamjiwnaang First Nation, Ontario', *Journal of Environmental Planning and Management*, 53:3 (2010), pp. 353–70.

6 Torny, 'Managing an Everlastingly Polluted World: Food Policies and Community Health Actions in the French West Indies'

1. In particular, this text owes a great deal to numerous discussions with Eric Godard, interministerial *chargé de mission* for chlordecone. I would also like to thank the members of the scientific committee on the chlordecone plan for their ever-pertinent questions and comments, and Matthieu Fintz and Nathalie Jas for their remarks on the preliminary versions of this text. I am very grateful to Chris Hinton for his translating assistance.
2. M. J. Beaugendre, *Rapport d'information de la Commission des Affaires Economiques, de l'Environnement et du Territoire sur l'utilisation du chlordécone et des autres pesticides dans l'agriculture martiniquaise et guadeloupéenne. Rapport n° 2430* (Paris: Assemblée Nationale, 2005), p. 91.
3. R. Carson, *Silent Spring*.
4. N. Jas in this book, on pp. 47–70.
5. Boudia, 'Global Regulation: Controlling and Accepting Radioactivity Risks'.
6. S. Boudia in this book, on pp. 71–88.
7. Jasanoff, *The Fifth Branch*.
8. B. L. Allen, 'Shifting Boundary Work: Issues and Tensions in Environmental Health Science in the Case of Grand Bois, Louisiana', *Science as Culture*, 13:4 (2004), pp. 429–48.
9. S. Krimsky, 'From Asilomar to Industrial Biotechnology: Risks, Reductionism and Regulation', *Science as Culture*, 14:4 (2005), pp. 309–23.
10. P. Jobin, 'Les cobayes portent plainte. Usages de l'épidémiologie dans deux affaires de maladies industrielles à Taiwan', *Politix*, 23:91 (2010), pp. 53–75.
11. B. L. Allen, *Uneasy Alchemy: Citizens and Experts in Louisiana's Chemical Corridor Disputes* (Cambridge, MA: MIT Press, 2003).
12. S. Frickel, 'Just Science? Organizing Scientist Activism in the US Environmental Justice Movement', *Science as Culture*, 13:4 (2004), pp. 449–69.
13. P. Brown, 'Popular Epidemiology Revisited', *Current Sociology*, 45:3 (1997), pp. 137–56; R. W. Clapp, 'Popular Epidemiology in Three Contaminated Communities', *The ANNALS of the American Academy of Political and Social Science*, 584:1 (2002), pp. 35–46.
14. S. Frickel and M. B. Vincent, 'Hurricane Katrina, Contamination, and the Unintended Organization of Ignorance', *Technology in Society*, 29:2 (2007), pp. 181–8.

15. F. Chateauraynaud, 'L'épreuve du tangible. Expériences de l'enquête et surgissement de la preuve', *Raisons pratiques*, 15 (2004), pp. 167–94.
16. This paragraph refers to the analyses made by M. Fintz, *L'autorisation du chlordécone en France 1968–1981* (Maisons-Alfort: AFSSET, 2009).
17. Beaugendre, *Rapport d'information de la Commission des Affaires Economiques, de l'Environnement et du Territoire sur l'utilisation du chlordécone et des autres pesticides dans l'agriculture martiniquaise et guadeloupéenne*, pp. 19–25.
18. H. Bonan and J.-L. Prime, *Rapport sur la présence de pesticides dans les eaux de consommation humaine en Guadeloupe, rapport IGAS n°2001-070 IGE n°01/007* (Paris: IGAS, 2001).
19. P. Balland, R. Mestre and M. Fagot, *Rapport sur l'évaluation des risques liés à l'utilisation de produits phytosanitaires en Guadeloupe et Martinique* (Paris: Ministère de l'Aménagement du Territoire et de l'Environnement/Ministère de l'Agriculture et de la Pêche, 1998), p. 47.
20. Balland, Mestre and Fagot, *Rapport sur l'évaluation des risques liés à l'utilisation de produits phytosanitaires*, p. 49.
21. We find just one quote on the substance: 'Cirad states it has detected chlordecone (forbidden in 93) at levels of several mg/L [*sic*] in the station's tap water'. Balland, Mestre and Fagot, *Rapport sur l'évaluation des risques liés à l'utilisation de produits phytosanitaires,* p. 50.
22. Bonan and Prime, *Rapport sur la présence de pesticides dans les eaux,* p. 47.
23. B. Latour, *Science in Action: How to Follow Scientists and Engineers through Society* (Cambridge, MA: Harvard University Press, 1987).
24. J. Snegaroff, 'Organochlorine Insecticide Residues in Soils and Rivers of Banana-growing Regions of Guadeloupe', *Phytiatrie & Phytopharmacie*, 26:4 (1977), pp. 251–67.
25. A. Kermarrec, *Niveau de la contamination des chaînes biologiques en Guadeloupe. Pesticides et métaux lourds 1979–1980. Rapport soumis au Ministère de l'agriculture,* 1980, at *www.antilles.inra.fr [accessed on 11 July 2012].*
26. F. Chateauraynaud and D. Torny, *Les sombres précurseurs.*
27. DSDS Martinique, *La qualité de l'eau potable* (Fort de France: DSDS de Martinique, 2008).
28. Bonan and Prime, *Rapport sur la présence de pesticides dans les eaux,* p. 35.
29. S. Bellec and E. Godard, *Contamination par les produits phytosanitaires organochlorés en Martinique: caractérisation de l'exposition des populations* (Fort de France: DSDS de Martinique, 2002).
30. S. Bellec and E. Godard, *Contamination par les produits phytosanitaires organochlorés en Martinique*, p. 8.
31. J. Desprats, J. Comte and G. Perian, *Cartographie du risque de pollution des sols de Martinique par les organochlorés. Rapport de phase 2* (Orléans: BRGM, 2003).
32. Y. Cabidoche, H. Jannoyer, H. Vannière and M. H. Loulergue, *Etat des connaissances sur la contamination des sols et des végétaux par le chlordécone aux Antilles françaises,* talk in GREPHY meeting, 17 October 2005.
33. M. Douglas, *Purity and Danger: An Analysis of Concepts of Pollution and Taboo* (London: Routledge and K. Paul, 1966).
34. Y. Cabidoche, C. Clermont Dauphin, R. Achard, A. Caron, P. Cattan, C. Chabrier, A. Lafont, M. Jannoyer and J. Sansoulet, *Stockage dans les sols et dissipation dans les eaux de la chlordécone, insecticide organochloré autrefois appliqué dans les bananeraies des Antilles Françaises* (INRA-CIRAD, 2006).

35. A. Pena-Vega, 'Les incertitudes de l'incident de Tchernobyl: L'émergence du sens dans les territoires contaminés en Belarus', *Sociétés,* 77 (2002) pp. 37–54.
36. Y. Cabidoche, H. Jannoyer and H. Vannière, *Conclusions du Groupe d'Etude et de Prospective 'Pollution par les organochlorés aux Antilles' Aspects agronomiques.* Rapport INRA-CIRAD, 2006.
37. 'Patates douces et toxiques durs', *Libération*, 16 October 2002.
38. AFSSA, *Avis de l'Agence française de sécurité sanitaire des aliments relatif à l'évaluation des risques liés à la consommation de denrées alimentaires contaminées par la chlordécone en Martinique et en Guadeloupe* (Maisons-Alfort: AFSSA, 2003).
39. B. Merle, V. Deschamps, S. Merle, A. Malon, A. Blateau, K. Pierre-Louis, P. Quénel and K. Castetbon, *Enquête sur la santé et les comportements alimentaires en Martinique (Escal 2003–2004); résultats du volet "consommations alimentaires et apports nutritionnels"*, (InVS, université Paris 13, CNAM, ORS Martinique, 2008).
40. AFSSA, *Avis de l'Agence française de sécurité sanitaire des aliments concernant deux projets d'arrêtés relatifs à la teneur maximale en chlordécone que doivent présenter certaines denrées d'origine végétale et d'origine animale pour être reconnues propres à la consommation humaine* (Maisons-Alfort: AFSSA, 2005).
41. AFSSA, *Avis de l'Agence française de sécurité sanitaire des aliments concernant deux projets d'arrêtés relatifs à la teneur maximale en chlordécone*, p. 13.
42. AFSSA, *Avis de l'Agence française de sécurité sanitaire des aliments concernant deux projets d'arrêtés relatifs à la teneur maximale en chlordécone*, p. 28.
43. AFSSA, *Première évaluation de l'exposition alimentaire de la population martiniquaise au chlordecone. Propositions de limites maximales provisoires de contamination dans les principaux aliments vecteurs* (Maisons-Alfort: AFSSA, 2005).
44. L. Boutrin and L. Confiant, *Chronique d'un empoisonnement annoncé. Le scandale du Chlordécone aux Antilles françaises 1972–2002* (Paris: L'Harmattan, 2007), pp. 152–7.
45. AFSSA, *Avis de l'Agence française de sécurité sanitaire des aliments relatif à la nécessité d'établir des recommandations particulières sur l'allaitement maternel au vu des bénéfices et des risques d'exposition au chlordécone pour les nourrissons martiniquais et guadeloupéens* (Maisons-Alfort: AFSSA, 2008).
46. Here we find an opposition analysed by Marc Barbier with regard to BSE: M. Barbier, 'Une interprétation de la constitution de l'ESB comme problème public européen', *Revue internationale de politique comparée*, 10:2 (2003), pp. 233–46.
47. AFP, 18 September 2007.
48. G. E. Markowitz and D. Rosner, *Deceit and Denial.*
49. D. Torny, 'L'administration des risques sanitaires face à l'éloignement de l'expertise: le cas français au tournant des années 2000', *Sociologies et société*, 39:1 (2007) pp. 181–96.
50. D. Pestre, 'The Historical Heritage of the 19th and 20th Centuries: Techno-science, Markets and Regulations in a Long-term Perspective', *History and Technology*, 23:4 (2007), pp. 407–20.
51. Frickel, 'Just Science?'
52. D. Fassin, 'Public Health as Culture. The Social Construction of the Childhood Lead Poisoning Epidemic in France', *British Medical Bulletin*, 69:1 (2004), pp. 167–77.
53. Public health officer, personal communication, October 2008.
54. The description of the JAFA programme and results is based on: J. Vincent, D. Camy, G. Thalmensi, M. Julien, M. Ledrans, P. Quénel, A. Blateau and E.Godard, 'Le programme de santé des Jardins Familiaux en Martinique. Bilan d'étape et perspectives', *BASAG*, 10 (2010), pp. 2–9.

55. S. Topçu, in this book on pp. 135–58.
56. Vincent, Camy, Thalmensi, Julien, Ledrans, Quénel, Blateau and Godard, 'Le programme de santé des Jardins Familiaux en Martinique. Bilan d'étape et perspectives'.
57. AFSSA, *Avis de l'Agence française de sécurité sanitaire des aliments relatif à l'actualisation des données scientifiques sur la toxicité du chlordécone en vue d'une éventuelle révision des limites tolérables d'exposition proposées par l'Afssa en 2003* (Maisons-Alfort: AFSSA, 2007).
58. InVS-Inserm, *Impact de l'utilisation du chlordécone aux Antilles françaises: Recommandations pour les recherches et les actions de santé publique à mettre en oeuvre* (Paris : InVS-Inserm, 2009).
59. Public Health officer, personal communication, October 2008.
60. D. Fassin, 'The Politics of Life: Beyond the Anthropology of Health' in F. Saillant and S. Genes (eds), *Medical Anthropology: Regional Perspectives and Shared Concerns* (Malden: Blackwell Publishing Ltd, 2006), pp. 253–66.
61. Chateauraynaud and Torny, *Les sombres précurseurs.*
62. Gillespie, Eva and Johnston, 'Carcinogenic Risk Assessment in the United States and Great Britain: The Case of Aldrin/Dieldrin'; Rothstein, Irwin, Yearley and McCarthy, 'Regulatory Science, Europeanization, and the Control of Agrochemicals'; R. Diggle, 'Regulatory Science and Uncertainty in the Risk Assessment of Pesticide Residues'.
63. D. Demortain, 'From drug crises to regulatory change'.
64. L. Multigner et al., 'Chlordecone Exposure and Risk of Prostate Cancer', *Journal of Clinical Oncology*, 28:21 (2010), pp. 3457–62.

7 Topçu, 'Chernobyl Empowerment? Exporting "Participatory Governance" to Contaminated Territories'

1. D. Pestre, 'Thirty Years of Science Studies: Knowledge, Society and the Political', *History and Technology*, 20:4 (2004), pp. 351–70; A. Irwin, 'The Politics of Talk: Coming to Terms with the "New" Scientific Governance', *Social Studies of Science*, 36:2 (2006), pp. 299–320.
2. C. Biewener and M.-H. Bacqué, 'Empowerment, développement et féminisme. Du projet de transformation sociale à la gestion néolibérale', in M.-H. Bacqué and Y. Sintomer (eds), *La démocratie participative. Histoire et généalogie* (Paris: La Découverte, 2011), pp. 82–3.
3. Irwin, 'The Politics of Talk'.
4. L. Levidow, 'European Public Participation as Risk Governance. Enhancing Democratic Accountability for Agbiotech Policy?', *East Asian Science, Technology and Society: An International Journal*, 1 (2007), pp. 19–51.
5. D. Pestre, 'Challenges for the Democratic Management of Technoscience: Governance, Participation and the Political Today', *Science as Culture*, 17:2 (2008), pp. 101–19; C. Thorpe and J. Gregory, 'Producing the Post-Fordist Public: The Political Economy of Public Engagement with Science', *Science as Culture*, 19:3 (2010), pp. 273–301.
6. P.-B. Joly and A. Kaufmann, 'Lost in Translation? The Need for "Upstream Engagement" with Nanotechnology on Trial', *Science as Culture*, 17:3 (2008), pp. 225–47.
7. B. Wynne, 'Public Engagement or Dialogue as a Means of Restoring Public Trust in Science? Hitting the Notes but Missing the Music', *Community Genetics*, 9:3 (2006), pp. 211–20.

8. F. Drake, 'Protesting Mobile Phone Masts: Risk, Neoliberalism, Governmentality', *Science, Technology & Human Values*, 36:4 (2011), pp. 522–48.
9. J. Clarke, 'New Labour's Citizens: Activated, Empowered, Responsibilized, Abandoned?', *Critical Social Policy*, 25:4 (2005), pp. 447–63.
10. Foucault, *The Birth of Biopolitics*.
11. M. Dean, *Governmentality*; T. Lemke, 'Foucault, Governmentality and Critique', *Rethinking Marxism*, 14:3 (2002), pp. 49–64.
12. N. Rose, 'Governing "Advanced" Liberal Democracies' in A. Barry, T. Osborne and N. Rose (eds), *Foucault and Political Reason: Liberalism, Neo-liberalism and Rationalities of Government* (London: UCL Press, 1996), pp. 37–64.
13. Thorpe and Gregory, 'Producing the Post-Fordist Public: The Political Economy of Public Engagement with Science'. Drake, 'Protesting Mobile Phone Masts'.
14. R. Frost, *Alternating Currents. Nationalized Power in France. 1946–1970* (Ithaca, NY & London: Cornell University Press, 1991); G. Hecht, *The Radiance of France*.
15. S. Topçu, 'L'agir contestataire à l'épreuve de l'atome'.
16. G. Medvedev, *The Truth about Chernobyl* (New York: Basic Books, 1991).
17. A. Yaroshinskaya, *Chernobyl, the Forbidden Truth* (Oxford: J. Carpenter, 1994); S. D. Schmid, 'Transformation Discourse: Nuclear Risk as a Strategic Tool in Late Soviet Politics of Expertise', *Science, Technology and Human Values*, 29:3 (2004), pp. 353–76.
18. B. Belbéoch and R. Belbéoch, *Tchernobyl, une catastrophe* (Paris: Allia, 1993).
19. A. Montaubrie, *La presse russe et la catastrophe de Tchernobyl (1986–1995). Représentations et mémoire* (Toulouse: GRHI-Université de Toulouse II, 1996).
20. G. A. Babcock, 'The Role of Public Interest Groups in Democratization. Soviet Environmental Groups and Energy Policy Making. 1985–1991' (PhD dissertation, Santa Monica Graduate School in Political Science, 1997). Available at: http://www.rand.org/pubs/rgs_dissertations/2006/RGSD135.pdf [accessed 24 September 2012].
21. G. Grandazzi, 'De Tchernobyl à la Hague. La vie quotidienne entre expérience de la catastrophe et épreuve de l'incertitude' (PhD dissertation, Caen University, 2004).
22. Yaroshinskaya, *Chernobyl, the Forbidden Truth*.
23. IAEA, *The International Chernobyl Project. Technical Report. Assessment of Radiological Consequences and Evaluation of Protective Measures. Report by an International Advisory Committee* (Vienna: IAEA, 1991).
24. WHO, *Report on the WHO Mission of Radiation Protection Experts to the USSR* (Geneva: WHO, 1989).
25. IAEA, *The International Chernobyl Project*, p. 511.
26. Grandazzi, 'De Tchernobyl à la Hague', p. 112.
27. P. Girard and G. Hériard Dubreuil, *Conditions de vie dans les territoires contaminés 8 ans après l'accident de Tchernobyl. Gestion du risque radiologique en Biélorussie: évaluation de la situation dans le district de Tchetchersk. Final Report within the Commission Programme for the Evaluation of the Consequences of the Chernobyl Accident* (Paris: Mutadis, 1995).
28. Grandazzi, 'De Tchernobyl à la Hague'.
29. Girard and Hériard Dubreuil, *Conditions de vie dans les territoires contaminés 8 ans après l'accident de Tchernobyl*.
30. V. B. Nesterenko, A.V. Nesterenko and A. Sudas, *Belarusian Experience in the Field of Radiation Monitoring and Radiation Protection of Population. Role of Governmental and Non-governmental Structures in Solving these Problems* (European Commission: Sage Project Deliverable, 2004).
31. Grandazzi, 'De Tchernobyl à la Hague', p. 113.

32. IAEA, *The International Chernobyl Project*, pp. 455–6.
33. IAEA, *The International Chernobyl Project*, p. 359 and p. 510.
34. IAEA, *One Decade After Chernobyl. Summing Up the Consequences of the Accident. Proceedings of an International Conference* (Vienna, 8–12 April 1996) (Vienna: IAEA, 1996).
35. Nuclear Energy Agency, *Chernobyl: Assessment of Radiological and Health Impacts. 2002 Update of Chernobyl: Ten Years On* (Paris: OECD, 2002).
36. IAEA, *The International Chernobyl Project*, p. 354.
37. P. Girard and G. Hériard Dubreuil, *Conséquences sociales et psychiques de l'accident de Tchernobyl. La situation en Ukraine sept ans après l'accident. Final Report within the Commission Programme for the Evaluation of the Consequences of the Chernobyl Accident* (Paris: Mutadis, 1994), p. 36.
38. Girard and Hériard Dubreuil, *Conséquences sociales et psychiques de l'accident de Tchernobyl.*
39. J. Lochard, 'Rehabilitation of Living Conditions in Territories Contaminated by the Chernobyl Accident: the Ethos Project', *Health Physics*, 93:5 (2007), pp. 522–26.
40. N. Rose and P. Miller, 'Political Power beyond the State: Problematics of Government', *The British Journal of Sociology*, 43:2 (1992), pp. 173–205.
41. Mutadis, *La réhabilitation des conditions de vie dans les territoires contaminés par l'accident de Tchernobyl. La contribution de l'approche Ethos, Proceedings of an International Seminar in Stolyn (Belarus)*, 15–16 November 2001.
42. G. Hériard Dubreuil, J. Lochard, P. Girard, J.-F Guyonnet, G. Le Cardinal, S. Lepicard, H. Ollagnon, V. Pupin, E. Rigby and T. Schneider, *The Ethos Project in Belarus. 1996–1998* (Paris: Mutadis, 2000), p. 26.
43. Mutadis, *Retour d'expérience de la gestion post-accidentelle de l'accident de Tchernobyl dans le contexte biélorusse. Rapport de synthèse* (Paris: Parex/ASN, 2007), pp. 16–17.
44. E. Hache, 'La responsabilité. Une technique de gouvernementalité néolibérale?', *Raisons politiques*, 4:28 (2007), pp. 49–65.
45. J. Lochard, 'Expertise et gestion des risques en matière nucléaire', *Revue française d'administration publique*, 3:103 (2002), pp. 471–81.
46. Author's interview with one of the coordinators of the Ethos Project, Fontenay-aux-Roses, 24 November 2008.
47. W. Krohn and P. Weingart, 'Commentary: Nuclear Power as a Social Experiment. European Political Fall-out from the Chernobyl Meltdown', *Science, Technology and Human Values*, 12:2 (1986), pp. 235–51.
48. J. E. Rigby, 'Principes et processus à l'œuvre dans un projet d'amélioration de conditions de vie dans les territoires contaminés par la catastrophe de Tchernobyl – ETHOS I (1996–1998)' (PhD dissertation, Université de Technologie de Compiègne, 2003).
49. Rigby, 'Principes et processus à l'œuvre dans un projet d'amélioration de conditions de vie dans les territoires contaminés par la catastrophe de Tchernobyl', pp. 189–205.
50. Nesterenko, Nesterenko and Sudas, *Belarusian Experience in the Field of Radiation Monitoring and Radiation Protection of Population*, p. 24.
51. UNDP and UNICEF, *The Human Consequences of the Chernobyl Nuclear Accident. A Strategy for Recovery.* A Report Commissioned by UNDP and UNICEF with the support of UN-OCHA and WHO, 2002, p. 77.
52. World Bank, *Belarus: Chernobyl Review* (Ukraine Belarus and Moldova Country Unit: Report n°23883 BY, 2002), p. 48.

53. A. Petryna, 'Biological Citizenship: The Science and Politics of Chernobyl-Exposed Populations', *Osiris*, 19:2 (2004), pp. 250–65, on p. 264.
54. UNDP, *Cooperation for Rehabilitation of Living Conditions in Chernobyl Affected Areas of Belarus. Annual Review* (Minsk: UNDP Support Project of the Core Programme), 2005.
55. I. Fiedler and G. Voigt, *Review of Infrastructures and Preparedness Systems in France, Germany and United Kingdom for Potential Releases of Radioactivity into the Environment* (European Commission: Sage Project Deliverable, 2004).
56. CEPN, *Strategies and Guidance for Establishing a Practical Radiation Protection Culture in Europe in case of Long-Term Radioactive Contamination after a Nuclear Accident: Final Report* (Paris: CEPN, 2005).
57. CEPN, *Guidance on Practical Radiation Protection for People Living in Long-Term Contaminated Territories. Sage Project Deliverable 5th Version* (Brussels: European Commission, 2007).
58. CEPN, *Strategies and Guidance for Establishing a Practical Radiation Protection Culture in Europe in case of Long-Term Radioactive Contamination after a Nuclear Accident: Final Report*, p. 22.
59. Mutadis, *Retour d'expérience de la gestion post-accidentelle de l'accident de Tchernobyl dans le contexte biélorusse*, p. 46.
60. Author's interview with a Codirpa member, Paris, 5 February 2009.
61. J. C. Niel and J. L. Godet, 'Construction d'une doctrine nationale pour la gestion de la phase post-accidentelle d'une situation d'urgence radiologique', *Contrôle*, 180 (2008), pp. 6–12.
62. Author's interview with an NGO delegate of Codirpa, Orsay, 3 March 2009.
63. S. Topçu, 'Confronting Nuclear Risks: Counter-expertise as Politics within the French Nuclear Energy Debate', *Nature and Culture*, 3:2 (2008), pp. 225–45.
64. W. Tchertkoff, *Le crime de Tchernobyl: le goulag nucléaire* (Arles: Actes Sud, 2006), p. 378.
65. Grandazzi, 'De Tchernobyl à la Hague'.
66. Nesterenko, Nesterenko and Sudas, *Belarusian Experience in the Field of Radiation Monitoring and Radiation Protection of Population*, p. 35.
67. Mutadis, *Retour d'expérience de la gestion post-accidentelle de l'accident de Tchernobyl dans le contexte biélorusse*, p. 69.
68. M. Fernex, 'Les mensonges clés. Ou comment effacer des mémoires les empreintes de Tchernobyl', *Trait d'union*, 22 (2002), pp. 18–22.
69. C. Castanier, 'Des structures écran au service du nucléaire', *Trait d'union*, 22 (2002), pp. 15–17.
70. A. Hirschmann, *Exit, Voice and Loyalty. Responses to Decline in Firms, Organisations and States* (Cambridge, MA: Harvard University Press, 1970).
71. R. Barbier, 'Quand le public prend ses distances avec la participation. Topiques de l'ironie ordinaire', *Natures Sciences Société*, 13 (2005), pp. 258–65.
72. Author's interview with an Acro member, Caen, 14 June 2006.
73. Author's interview with an Acro member, Paris, 5 February 2009.
74. Lemke, 'Foucault, Governmentality and Critique'.

INDEX

For Product Safety Concerns and Information please contact our EU representative GPSR@taylorandfrancis.com
Taylor & Francis Verlag GmbH, Kaufingerstraße 24, 80331 München, Germany

www.ingramcontent.com/pod-product-compliance
Ingram Content Group UK Ltd.
Pitfield, Milton Keynes, MK11 3LW, UK
UKHW011012180926
475701UK00004B/702

9781138664760